T0343989

43 VISIONS FOR COMPLEXITY

published by

World Scientific Publishing Co. Pte. Ltd.
5 Toh Tuck Link, Singapore 596224
USA office: 27 Warren Street, Suite 401-402, Hackensack, NJ 07601
UK office: 57 Shelton Street, Covent Garden, London WC2H 9HE

British Library Cataloguing-in-Publication Data
A catalogue record for this book is available from the British Library.

Cover: Olaf Osten, *Tele-Vision I-168*, Acrylics on map, 2014,
Endpaper: Olaf Osten, *Tele-Vision I-50*, Oil on canvas, 2004

Exploring Complexity – Vol. 3
43 VISIONS FOR COMPLEXITY

ISBN 978-981-3206-84-7

Printed in Singapore

43 VISIONS FOR COMPLEXITY

Stefan Thurner
editor

NEW JERSEY · LONDON · SINGAPORE · BEIJING · SHANGHAI · HONG KONG · TAIPEI · CHENNAI

On the occasion of the opening of the Complexity Science Hub Vienna on 23 May 2016 some fifty leading complexity scientists from all over the globe were invited to share their visions. They were asked where they see the present frontiers of the science of complex systems, where complexity science will most likely be in a decade from now, and in what ways it can contribute to address some of the big challenges that our societies and the planet are currently facing. Their visions have been collected in this book.

A general theme that percolates throughout these visions is the transformative power of data. They do not only impact the world around us at a tremendous pace but also science, this great human achievement of understanding the natural and the social world in a rational way. The bottleneck to understanding is no longer the availability of data. Rather, it is how to make sense of data in quantitative and predictive ways. The science of complex systems is doing exactly this: it provides new methods and novel ways of fundamentally understanding systems that were thought to be unintelligible only a few decades ago. Complexity science links expertise of state-of-the-art mathematics, modelling, data and computer science with specific questions posed from various disciplines, such as on future forms of data-driven medicine, production processes and the future of the economy, systemic risks, and the nature of innovation and creativity.

Disciplinary boundaries are rapidly dissolving. This does not mean that disciplinary training of scientists has become obsolete. It means that scientists are able to cover at least two areas: their own area of disciplinary expertise while the other is the mastery of the common language between disciplines, namely applied mathematics and the ability of computational modeling. By establishing and reinforcing this common language the science of complex systems is a discipline without disciplinary boundaries.

The Complexity Science Hub Vienna is a new collaborative institution with the aim to become a focal point of complexity

science in Europe. The founding partners are four universities – TU Wien, Medical University of Vienna, Graz University of Technology, Vienna University of Business and Economics – and two research institutions – the AIT Austrian Institute of Technology and the International Institute for Applied Systems Analysis. Its guiding principle and structure is simple. Each partner institution contributes by appointing two local faculty researchers; the Hub hires catalytic researchers and, in addition, sponsors an external faculty and visitor program. The aim is to provide at any point in time a creative and scientifically thriving environment that is free of bureaucratic constraints. Public outreach activities will be designed to share the excitement of complexity science with a wider audience. The Hub is also a node in a network of international partner institutions, including the Santa Fe Institute, the Complexity Institute at Nanyang Technological University in Singapore, Arizona State University, and the newly founded Institute for Advanced Study at the University of Amsterdam. Through this network a lively exchange of scientists, question posers, students, and postdocs is envisioned, enabling the most important exchange – that of ideas.

The activities of the Hub are supported by the Austrian National Foundation and the Austrian Federal Ministry for Transport, Innovation, and Technology. Additional funding will be sought at national and European level. The scientific quality is ensured by the eminent members of a Science Advisory Board chaired by Helga Nowotny.

The aims for the Hub will be reached when it becomes possible to create physical and virtual platforms for the most creative, talented, non-dogmatic, and open-minded scientists to interact on the most pressing and most fascinating questions in science today. They all belong to and constitute the realm of complex systems.

Stefan Thurner
Vienna, September 2016

On a summer day in 2005 Jan Vasbinder visited me at the European Science Foundation. Jan had a dream: to set up a European institute dedicated to sciences for our future and based on the principles of the Santa Fe Institute. It was to be an independent institute that would grow a new kind of scientific research community, one emphasizing multi-disciplinary collaboration in pursuit of understanding the common themes that arise in natural, artificial, and social systems. It was a wonderful dream, and I bought into it right away. So I became founding father and initial funder of Institute Para Limes IPL. In March 2006 ESF hosted the first meeting of the founding fathers, in which the structure, governance and the first research agenda were decided upon.

IPL had a very good start, and fared well until the end of 2008, when the financial crisis hit and the committed funds evaporated. That crisis put an end to the plan to house IPL in a 14th century monastery in the Netherlands that was being reconstructed for that purpose. But Jan and I did not give up. By that time I had become the provost at Nanyang Technological University in Singapore. From there Jan and I looked for a suitable location for IPL in Europe. We had intensive discussions with the municipal leadership in Strasbourg and Lindau. But both cities lacked the drive to make our dream come true. Then we hit on Vienna.

In Vienna everything came together: Helga Nowotny, the grand lady of European science who embraced the idea and put her enormous drive and resourcefulness into it, Stefan Thurner, external faculty member of SFI and a man with a vision and a research agenda, four universities that provide a stable base for funding and now a real palace. Some adaptations had to be made to realize our dream within the Viennese context, but never have the chances been better, and I congratulate everyone who made this happen.

» Inevitability of Interdisciplinary Approaches «

My vision is that complexity is a defining feature of all grand challenges that humanity has to meet and of almost all problems that present day science is addressing or needs to address. Therefore, all universities should have a complexity program that students of all disciplines should follow. That program should familiarize students with ways of looking at objects of study from the bottom up, seeing them as systems of interacting elements that form, change, and evolve over time and it should introduce them to the concepts and tools that complexity science provides. Such programs must not position complexity science as a new discipline. Instead it must stress the inevitability of interdisciplinary approaches to recognize and address the complexity of the real world. After all, preparing students to function in the real world is one of the main responsibilities of universities.

NTU takes this responsibility very seriously. Five years ago we started a complexity program. Since April 2014 we have a complexity institute that focuses on research on essential problems in complexity and we have Para Limes, the follow up of the initial complexity program.

I believe the Complexity Science Hub in Vienna can play a leading role in Europe in bringing home the message to universities that understanding complexity should be an essential part of any university curriculum.

Bertil Andersson / Nanyang Technological University, Singapore
is a world-renowned plant biochemist, who has pioneered research on the artificial leaf, which includes authoring more than 300 papers in basic photosynthesis. Since becoming President of NTU Singapore in 2011, Professor Andersson has led the University to global distinction. NTU is the world's fastest-rising young university and is ranked first among the world's young universities. Before joining NTU as its Provost in 2007, he was Chief Executive of the European Science Foundation. Professor Andersson has had a longstanding association with the Nobel Foundation. He was Chairman of the Nobel Committee for Chemistry as well as a Trustee of the Nobel Foundation.

We now have had 400 years of extremely successful reductionist science in which systems have been understood at finer and finer levels. We have also had 40 or so years of looking in the other direction, at how system behavior emerges from the interactions of its lower-level elements. This new movement in science, complexity, is not old and we stand very much at the beginnings of what it will bring.

In no small way complexity has come out of the arrival of computers. Before computers, if we wanted to understand systems, we had to treat them largely as linear, in stasis or equilibrium, predictable, and expressible in equations. Now, with the help of computers, we can look at systems that are nonlinear, not in equilibrium, not predictable, and expressible in algorithms. These latter properties don't always look familiar to us or even scientific or well-ordered, but they are opening to us a world that is vast and unknown. In mathematics, physics, biology, earth sciences, and economics we are seeing structures that are strange and wonderful and unexpected, and that brim with messy vitality.

» COMPLEXITY IS PRACTICAL IN ITS INSIGHTS «

One of the joys of complexity is that we can use it to probe statistically the huge quantities of data that real-world systems, whether medical or commercial or logistical, are delivering via sensors and the internet. We are thus finding new insights into real world systems. Another of the joys is that we can create computer-based models in which individual elements – people, cars, cells in the immune system, financial instruments, government policies – can interact and we can directly observe the results. Complexity is practical in its insights, and again we are at the beginnings of this type of guidance from science.

In the past, Austria, with Vienna in particular, has been a great leader in all the sciences. So it is entirely appropriate that this new science of complexity be instituted and nourished in Vienna. I am confident the Complexity Science Hub Vienna will flourish and I wish it every success. Alles Gute!

W. Brian Arthur / Santa Fe Institute
is best known for his theoretical work on increasing returns or positive feedbacks in the economy and their role in locking markets in to the domination of one or two players. He is one of the pioneers of the science of complexity and one of the founders of the Santa Fe Institute, where he served many years on its Science Board and Board of Trustees. Brian Arthur is the recipient of the inaugural Lagrange Prize in Complexity Science in 2008 and the Schumpeter Prize in Economics in 1990. He has been dean and Virginia Morrison Professor of Economics and Population Studies at Stanford, and Citibank professor at the Santa Fe Institute.

Complex systems in their environments are open, emerging and dying, individuated and heterogeneous, multilevel organized and controlling from inside their boundary conditions. The phenomenological and theoretical reconstruction of their multi-level dynamics is the main epistemological challenge of complex systems science.

The phenomenological reconstruction is starting with shared experimental protocols for cohorts of individuated 3D+t multimodal multiscale dynamics. The individuated data is noised and some parts are missing. This reconstruction provides an augmented phenomenology, identifying at each organizational level the components, their relations, and their hyper-symbolic dynamics. Such hyper-symbolic dynamics is a generalization of the symbolic dynamics. It is obtained i) by multimodal 3D+t unsupervised deep learning for prototyping and categorizing components and relations and strongly based on algebraic topology and geometry, ii) by supervised deep learning of the names of components and relations by scientists and experts. It is functionally analog to what is done by the very young child categorizing the components and relations of the world around and then acquiring the names of the categories with the parents.

» THE PREDICTION PARADIGM HAS TO SHIFT «

The criterion of success of phenomenological reconstruction of noised and missing data is based on information geometry. The theoretical reconstruction needs to generalize the theoretical reconstruction in physics and statistical mechanics for complex systems. The prediction paradigm has to shift from predicting what is happening in certainty to what can happen in probability. At each organizational level of complex systems, the projection theorem of statistical mechanics has to be extended to provide generalized Fokker-Planck equations taking into account their whole history. The statistical field theory for binary relations has to be extended into a multinomial statistical field theory for n-ary relations. The theoretical reconstruction has to adapt its little number of parameters for each prototype when useful for recommendations in prevention or resilience of a specific complex system categorized by this prototype. The criterion of success has to compare the empirical probability distribution of the phenomenological reconstruction with the theoretical probability distribution. It is also based on information geometry.

Both the phenomenological and theoretical reconstruction of multilevel dynamics are using information geometry as criterion of success.

Paul Bourgine / French National Center for Scientific Research, Paris
is president of the Complex Systems Digital Campus (CS-DC UNESCO UniTwin), honorary director of RNSC (National Network of Complex Systems), and former director of CREA-Ecole Polytechnique. He holds a diploma of the Ecole Polytechnique (1968), a PhD in Economics (1983), and a habilitation in cognitive science (1989). Paul served as the first president of the Complex Systems Society. Scientific interests include complex adaptive systems and large interactive networks. Current research fields are genetic networks, neural networks, social networks and social cognition, learning and co-evolutionary dynamics. He served as co-chair of many international conferences, among other, the first European Conference on Complex Systems (2005), and the first World e-Conference CS-DC'15.

As society becomes more and more interconnected (we just started to add our environment to the net with the Internet of Things revolution), the net will be the place where most interactions will take place.

The mathematical instruments to describe, model, and possibly improve the various systems are developing just now (multi-level complex systems, multiplex of networks, etc.). It is likely that in such a structure we shall tend to avoid centralized systems of control. In the future it will become crucial to allow people to build their own online reputation and to recognize the reputation of others without certification from a central authority. The work of the next years will be to develop the conceptual, theoretical, and technological tools necessary to distribute access to resources and create new social, economic, and financial

» The Net will be the place where most interactions will take place «

applications allowing people to have better access to opportunities and improved possibilities to exercise their rights as citizens and investors.

The most ambitious possibility opened by decentralized digital services is giving users back the control over their data. The present model in which users exchange personal data for services provided by companies has several drawbacks. First, such exchange is often implicit, with very few users understanding the full implications of it. Second, it is largely illiquid, in the sense that, once a user has surrendered her data to a company, she has no choice of trading it with other companies, precisely because the data belongs to a unique (central) entity. An alternative model consists in users retaining full ownership of their data, simply notifying a provider of their intention to trade them in exchange for money or services.

Guido Caldarelli / IMT Institute for Advanced Studies, Lucca
received his PhD from SISSA, after which he was a postdoc in the Department of Physics and School of Biology, University of Manchester. He then worked at the Theory of Condensed Matter Group, University of Cambridge, where he collaborated with Robin Ball and stayed in Wolfson College. He returned to Italy as a lecturer at the National Institute for Condensed Matter (INFM) and later as primo ricercatore in the Institute of Complex Systems of the National Research Council of Italy. In this period he was also the coordinator of the Networks subproject, part of the Complexity Project, for the Fermi Centre. He also spent some terms at University of Fribourg, Switzerland. In 2006 he has been visiting professor at École Normale Supérieure in Paris.

The basic goal of the sciences is to point to, and explain, emergent phenomena: what we would not have guessed given what we knew before. This lack of predictability can come from a change of scale (more is different; physics), a change of descriptive language (lost in translation; the human sciences), or just patience on the part of the observer (self-organization; biology). Nothing worth knowing can be predicted.

But at the frontier of machine learning, and in the databases of the modern corporation, it is a simple fact: nature, even (to take an example) human behavior, is extremely predictable. Algorithms will only improve their already remarkable accuracy in guessing what we will eat, how we will sleep, and what we will say when we wake up the next morning.

For science to thrive in this era, it needs to become ever more explicit, ever more rigorous, about its true subject. Our challenge is to build a better theory of the unpredictable. Not (of course) a theory of noise, but a theory of what is unexpected, unpredicted and yet patterned: a theory, for lack of a better word, of creativity.

» HUMAN BEHAVIOR IS EXTREMELY PREDICTABLE «

Most theories of creativity we have now are depressingly inadequate. They cash out creativity as nothing more than combinatorics: novel juxtaposition or innovative analogy-making. A small minority of theories invoke Gödelian accounts and connect the incompleteness of logical systems to the unpredictable behavior of feedback loops. Despite their opacity, members of this latter group have the advantage of experience: the loops they predict do appear necessary for complex behavior. But even our best theories today are not nearly enough.

Good theories of creativity will tell us how to measure it; will describe and explain its taxonomy; will point to its origins and causal correlates. Creativity lacks a Darwin, but also a Carl Linnaeus; a Tycho Brahe just as much as a Galileo. Without them, science will increasingly confuse successful predictions for advances in knowledge.

In 1956, the cyberneticist Norbert Weiner warned of the *inhuman use of human beings.* Sixty years later, we are predictable automata in the eyes of our own machines. We are in critical need of guides through our terabytes – guides that can point out, and help us value, what is surprising about ourselves.

Simon DeDeo / Carnegie Mellon University Pittsburgh, Santa Fe Institute
is external professor at the Santa Fe Institute, and assistant professor at Carnegie Mellon University, where he runs the Laboratory for Social Minds. The laboratory has studied a diverse set of systems that includes the courtrooms of Victorian London and the parliaments of the French Revolution, the research strategies of Charles Darwin, self-governance in Wikipedia, colluding petrol stations in the American Midwest, conflict and power in the social avians.

Complexity science and big data science are challenging areas that require competences and inputs from different fields. These areas are interdisciplinary by their nature and problems and challenges that they are addressing are intrinsically related to the continuous evolution of our society.

We are witnessing interesting times rich of information, readily available for us all. Using, understanding, and filtering such information have become a major activity across science, industry and society at large. Our society has become a global information processing system where news propagate and impact on the real economy at increasingly fast rates with increasingly large effects. It is therefore important to have tools that can analyze this information while it is generated and that can provide ways to reduce complexity and dimensionality while keeping the integrity of the dataset. This information folding poses important challenges to science and industry: i) data gathering by large-scale experiments; ii) data cleaning and data handling including sensitive data and related ethical problems; iii) the managing of huge databases which are continuously updated; iv) the development of models that can account for the continuous change of input information; v) the development of algorithms that can handle the computational complexity.

In the next decade complexity science and big data will grow more and more and will require diverse competences of an increasingly large number of researchers with different

» Society has become a global information processing system «

backgrounds. This is indeed the big challenge: to create places where ideas and knowledge can be exchanged and a common language developed. This hub in Vienna is the right place at the right time.

For what concerns economics and finance – my own specialization – this hub can greatly advance research in risk management, systemic risk, data analytics, and regulation policies and it will pave the way for a new generation of forecasting tools, useful for regulators and industry, conceived to deal with large datasets comprising entire markets, different assets, news feeds, and other information sources. These tools will help to stabilize financial fluctuations and will provide new tools to regulators.

Complexity science, big data science, and now the Complexity Science Hub Vienna are helping society to propose, study, and develop new approaches to prediction, forecasting, and risk in systems that are intrinsically affected by uncertainty, noise, and extreme events, making – in Niels Bohr's words – *prediction very difficult, especially about the future.*

Tiziana Di Matteo / King's College London

A trained physicist, Tiziana Di Matteo took her degree and PhD from the University of Salerno in Italy before assuming research roles at universities in Australia and Britain. She works in the Department of Mathematics at King's College London in econophysics, complex networks, and data science. She has authored over 90 papers and gave invited and keynote talks at major international conferences in the US, across Europe, and Asia, making her one of the world's leaders in this field. She is co-editor in chief for the Journal of Network Theory in Finance and editor of the European Physical Journal B, editor of Quantitative Finance Letters, and guest editor of several other volumes. She has been consultant for the Financial Services Authority and several hedge funds.

From my personal point of view, complexity is the science of integration: integrating different mechanisms, scales, and dynamical rules to generate a new and emergent system. It integrates other traditional scientific disciplines in a common framework. Complexity science makes use of powerful tools of statistical physics, mathematics, and computer science, to bring back answers and even predictions to the originating disciplines.
In particular, networks provide the topological skeleton through which interactions take place.

» COMPLEXITY IS THE SCIENCE OF INTEGRATION «

Complex networks have different degrees of inherent complexity. They are dynamical, weighted, hierarchical, multiscale, multilayer, multi-featured, etc. It will be fascinating to witness how these different facets of the topological structure will help us to understand, model, control, and even predict the behavior of the most complex systems, as those formed by the whole human kind. But the roots of complex systems are to be searched in much smaller scales. Integrating them out will help us in understanding the whole system.

Albert Diaz-Guilera, complexitat.cat; Universitat de Barcelona
received his degree in physics at the Universitat de Barcelona and his PhD in science at the Universitat Autonoma de Barcelona, followed by postdoctoral stays at the Gorlaeus Laboratories, Leiden, and the Centre de Physique de Solide in Sherbrooke, Canada. He is the coordinator of the Catalan Network for the Study of Complex Systems (complexitat.cat). Currently his research is focused on general aspects of complexity, particularly on complex networks. By education a statistical physicist, his research has been broadening to cover aspects in many different fields: biology, economy, social sciences, computer science, and linguistics. Albert is author of more than 100 articles in physics and interdisciplinary journals.

Pierre Teilhard de Chardin referred to the fusion of biological life, human culture, and technology as the noosphere. Technological improvement is causing the noosphere to evolve rapidly, driving the enormous increase in human population over the last 10,000 years and the transformation (and devastation) of the biosphere. The rapid proliferation of the internet is changing human culture, including everything from the way we find mates to the way democracy functions, or fails to function. The emergence of the BINC (Bio, Info, Nano, Cogno) technologies promises to further accelerate this change. We are acquiring an ever-increasing ability to engineer devices at a molecular level, to control the genome, and to create new forms of life and intelligence.

While we might have the ability to alter our world as we wish to suit our needs, arriving at collective action to act on this ability remains elusive. This is in part because our understanding of our collective behavior is so poor. We have a relatively good understanding of the forces driving climate change, for example, but a poor understanding of the forces that shape our economy and society, and how to harness these forces productively. We have an increasing need to model ourselves. Just as a human's model of itself is called consciousness, society's models of itself might be called collective consciousness. We are undoubtedly

» We have an increasing need to model ourselves «

a complex system and we need to build models to enhance our collective consciousness accordingly; unfortunately many fields in social science, such as economics, are still employing 20th century concepts and methods to solve 21st century problems. The problems we face require complex systems engineering, which involves both understanding the dynamics of the noosphere and techniques for controlling it.

Fortunately modern technology gives us an unprecedented capability to collect data about ourselves and model ourselves, to create a science of complex systems engineering for understanding all aspects of the noosphere. Our ever-increasing computational power makes it possible to build sophisticated models of our behavior that go far beyond anything we could ever do using the primitive technology of closed form mathematics. Complex systems engineering is more like gardening than traditional design, in that one has limited control of the details; rather all we can hope to do is to tinker with the rules and let the consequences evolve. This creates an enormous opportunity for complex systems to prove itself by solving practical problems.

The world is our garden and we need to tend it.

J. Doyne Farmer / University of Oxford; Santa Fe Institute
is director of the Complexity Economics Program at the Institute for New Economic Thinking at the Oxford Martin School, professor in the Mathematical Institute at the University of Oxford, and an external professor at the Santa Fe Institute. His current research is in economics, including agent-based modeling, financial instability, and technological progress. He was a founder of Prediction Company, a quantitative automated trading firm that was sold to the United Bank of Switzerland in 2006. His past research includes complex systems, dynamical systems theory, time series analysis, and theoretical biology. During the eighties he was an Oppenheimer Fellow and the founder of the Complex Systems Group at Los Alamos National Laboratory.

First, I believe we're entering an era of synthesis of modeling. Over the past 20 years, we've seen the proliferation of many isolated complex systems models. I think we now need tools for researchers, policy makers and the public to share models. Sharing could happen by stacking different layers of spatial agent-based models in geographic information systems and projecting interactive visualization out onto shared surfaces. Further, we need to make model authoring tools much more accessible to the point where motivated policy makers can author on their own. With the increased ability to author and share models, I believe this will allow us to scale our research to understand and manage the many interacting systems that make up our complex world.

Second, I'd like to see CSH venture into risky research and go after bold questions. For example, I'd like to see a pursuit of a

» Entering an Era of Synthesis of Modeling «

universal theory of complex systems and/or pursue formal models of living systems. In particular, I'd like a theory that allows one to define how far-from-equilibrium a given agent-based model is and how its action gradients are imported into the constraints of the local structure that emerge to dissipate these gradients. Both the action gradients and structural constraints are measurable in bits. While I don't presume to know how this will be accomplished, I suspect ideas leveraging symmetry, action, and duality will be useful.

I am excited to see the opening of the Complexity Science Hub Vienna – I hope over the next 10 years the Hub will help develop these two main areas in complexity science. I look forward to CSH's role in developing applied complexity research.

Stephen Guerin / Simtable LLC
is CEO of Simtable and the principal behind RedfishGroup. He lectures as a faculty member of Santa Fe Institute's Complex System Summer School. Simtable produces interactive simulations for firefighters and emergency managers that are projected onto physical 3D sandtables. RedfishGroup takes on R&D projects in the areas of agent-based modeling, human computer interaction, and ambient computing (spatial augmented reality). Stephen formed RedfishGroup in 1991 initially focusing on graphics applications in commercial prepress and interactive media. With the emergence of the web in the mid 90's Redfish consulted in Beijing and Shanghai to multinational companies, Chinese ministries and the US Embassy in China. In the late 90s, Stephen focused his research in Cognitive Science with applications to distributed software systems and artificial life.

With all the big data available, would it be possible to build a crystal ball allowing us to see everything that is going on in the world in real time? Such projects are in fact under way, built by the military and research centers around the world. These are also political projects, because *knowledge is power*. Will so much data enable the ruling of a *wise king* or a *benevolent dictator*? Could society even be run like a giant machine? Indeed, there are companies that work on such concepts, for example, Google and IBM. They aim at building an operating system for our society that would try to steer our decision-making and behavior with personalized information.

This brings our society at a crossroads. We will certainly live in a data-based society – but what kind of society will it be? Some voices claim that democracy is an outdated technology. In fact, we see that democracy is already in trouble in several countries. We might easily lose what we have built over hundreds of years – freedom, human dignity, justice, pluralism, democracy, culture, etc.

In many cases, the magic formula *more data = more knowledge = more power = more success* does not work. Correlation does not equal causation. There is also a technical reason for this: even though processing power increases exponentially, data volume increases even faster. Therefore, the fraction of data we can process is decreasing over time.

Moreover, as we go on networking the world, systemic complexity is growing even faster, which implies a loss of top-down control and a need for distributed control. Hence, we need to build an upgraded, digital democracy: *democracy 2.0*. For this we need to learn how to bring the knowledge of people and artificial intelligence systems together. This takes online deliberation platforms, because it is often not the best individual solution that wins, but a combination of diverse solutions.

» We need to build an upgraded, digital democracy «

We also need a new kind of economic system – capitalism 2.0 – which is liberal, democratic, participatory, social, and ecological. It would measure, value, and trade externalities. Furthermore, it would include a multi-dimensional incentive and reward system: *finance 4.0* or *social-ecological* finance that would allow us to create feedback loops in the system in order to support favorable kinds of self-organization. In such a way a circular economy would result, as people measure the environment in a crowd-sourced way and share the data with everyone. Doing so citizens would create and earn different kinds of money bottom-up, which would then rise to the top, benefitting everyone. Such a system can now be built by combining the Internet of Things, block chain technology, and complexity science. It would mitigate major problems challenging us such as the loss of classical jobs, insufficient taxes, climate change, and resource shortages.

With this in mind, we have started to build the *nervousnet* platform: a *planetary nervous system* based on Internet of Things technology run by citizens, using smartphones as sensors to measure the environment. Our goal is to develop principles and technological solutions that would help to turn the digital desert in Europe into a diverse digital rainforest with digital opportunities for everyone. Requiring sufficient interoperability, existing products and services could then be combined to create new products and services, thereby unleashing combinatorial innovation.

Let's do this together now!

Dirk Helbing / ETH Zurich
is professor of computational social science at the Department of Humanities, Social and Political Sciences at ETH Zurich. He is coordinator of the FuturICT Initiative and he is an elected member of the prestigious German Academy of Sciences Leopoldina. In 2014 he received an honorary PhD from Delft University of Technology. He is affiliate professor at the Faculty of Technology, Policy and Management at TU Delft, where he leads the PhD school in Engineering Social Technologies for a Responsible Digital Future.

The world consists of inter-connected processes. It is an illusion to think that *things* exist. As argued by Alfred North Whitehead processes are ontologically fundamental. Since the building blocks do not consist of *things* with specific intrinsic properties, we can hope that a general science of the processes underlying and controlling the behavior of apparently very different situations, such as the evolution of an ecosystem or the performance of a piece of music by a band, may very well exist.

A saxophone and a tree don't have much in common. But a jazz band and a forest might very well have. Say for instance in the way information is moving around amongst the components and in how structures evolve. The similarity may extend further. For instance, a saxophone is of course not really a *thing*. It is a collection of processes that moves objects around and generates the process of variation in air pressure, which in our mind becomes the wonderful phenomenon we call music. Similar for ecosystems. What we see as *things*, building blocks, or components at one level are at another level themselves collections of components participating in processes.

Why is this rather self-evident observation important to us? Because as soon as one realizes that the world is made of inter-connected processes and not of *things*, one immediately realizes why complexity science is the most fundamental of the sciences and why a place like the Complexity Science Hub Vienna is likely to create more fundamental insights about the world we are surrounded by than, say, the Large Hadron Collider ever will do. How should complexity science go about uncovering the generalities behind the super complex systems such as the brain, the economy, or ecologies? By finding ways to extend the

»It is an illusion to think that ›things‹ exist«

tremendous successes of statistical mechanics. The science of multi-component systems developed around 1900 not least here in Vienna by Boltzmann. At present, statistical mechanics has its greatest successes when applied to equilibrium systems. Complexity science needs to discover ways to systematically extend statistical descriptions too far from equilibrium. In doing so we must not be naïve, we must not be too attached to our old formalisms. But there are indications that ordinary statistical mechanics of systems at or near equilibrium may point us in the right direction. We need to expand our understanding of intermittency, tipping points, stability – in short: we need to understand the emergence of hierarchies. We need to identify which aspects of complex systems are typical across many types of systems and we need to learn what kind of mathematical formalism is able to capture and describe the emergent hierarchical structures.

Big data will undoubtedly be a big help – but only if we use the access to big data to identify simple unifying concepts. We won't gain much insight by making our models as complicated and involved as the phenomena we try to understand or as data rich as our big data banks. We need understanding of co-evolution, simple networks, statistical mechanics, and information theory.

Henrik Jeldtoft Jensen / Imperial College London
is a professor of mathematical physics and leader of the Centre for Complexity Science at Imperial College London. He works on the statistical mechanics of complex systems. He has worked on the dynamical properties of condensed matter systems and developed the tangled nature *model of evolving ecosystems, which is currently used to develop the* tangled finance *approach. His two books on complexity science* Self-organized Criticality *and* Stochastic Dynamics of Complex Systems *(with Paolo Sibani) have attracted very broad interest. Henrik Jensen has more recently worked on brain dynamics and structure by analyzing fMRI and EEG data, beside of various research projects in evolutionary ecology and finance and economics.*

There really is no *science of complexity*. Rather we have a fairly well developed set of tools to examine diverse complex and complex adaptive systems. These tools include now familiar ideas of nonlinear dynamical systems, bifurcation theory, and stochastic models, as well as agent-based models such as BOIDS. These tools have been well developed in the past 30 years and we are now underway with the applications of such tools. As B. Arthur noted in analogy, the railways in Britain caused a surge in their stock values, which then fell as the bubble burst, but most of the track was laid afterwards. So, too, complexity burst upon the scene in the late 1980s, largely at the Santa Fe Institute. If that messianic era is now, naturally, past, we are enabled to lay enormous tracks as we proceed.

But also there are new frontiers. At the risk of noting my own work with G. Longo, we believe we have shown that *no laws entail the evolution of the biosphere.* If so, since the biosphere is part of the universe, strong reductionism must fail. More, the becoming of the biosphere, without selection achieving it, enables the new opportunities for its further evolution in a typically *unprestatable* way. Not only do we not know what will happen, as in flipping a fair coin, we do not even know what can happen. The biosphere becomes into this *unprestatable* but ever newly opening set of possibilities into its *adjacent possibles.*

So, too, does the global economy become. Both the biosphere and the economy have exploded in diversity, one in the past 3.7 billion years, the other in the past 100,000 years. That which already exists provides the very *context* that affords yet new opportunities that evolution and entrepreneurship can *seize*, yielding new entities in the enlarged *contexts*. The *adjacent*

» THE ›ADJACENT POSSIBLE‹ IS RELATIONAL «

possible is relational: The current *context* and the very entities that are part of the *context* that can seize the opportunities afforded by their context. These *entities* are organisms and entrepreneurs in the becoming biosphere or economy. The new actuals that arise, organisms with swim bladder derived from the lungs of lung fish by Darwinian preadaptation, or the *web*, for example, provide the new, typically enlarged, contexts in which more evolution, organisms, and enterprises in the biosphere and economy can arise. Google and Amazon live on the *web* but could not have done so 100 years ago. The more the diversity of the *context*, the more opportunities exist. New organisms and enterprises, given their contexts, come into existence, thence providing an ever enlarging set of opportunities for yet more new entities to come into existence and enable, not cause, yet more opportunities.

I do not think we can *mathematize* this becoming. If not, we are beyond Pythagoras, and confront a new frontier of the sciences of complexity.

The Complexity Science Hub Vienna has the fine opportunity to further what may now be called the standard theory of the sciences of complex systems. In addition, and ineluctably, CSH has the possibility of going beyond the standard theory.

Stuart Kauffman / Institute for Systems Biology, Seattle
was educated at Dartmouth College, Oxford, and University of California Medical School, receiving his M.D in 1968. At that time he developed Random Boolean Networks *as a first model of complex genetic regulatory networks, and proposed that cell types correspond to dynamical attractors in such networks. In 1971 he proposed that molecular reproduction arises due to the spontaneous formation of collectively autocatalytic set. He has published five books:* The Origins of Order, At Home in the Universe, Investigations, Reinventing the Sacred, *and* Humanity in a Creative Universe. *His honors include a MacArthur Fellowship, the Gold Medal of the* Accademia dei Lincei, *and a fellowship of the Royal Society of Canada.*

We live in a world of demographic explosion, bloody ethnic and religious wars, migration galore, increasing inequality and crime, global financial crises, and the danger of pandemics.

Changes of human society have accelerated to an unprecedented extent challenging our ability to adapt. Technological development has opened entirely new channels of communication, induced new behavioral patterns, influenced substantial organization principles and its products are becoming history-forming factors.

Society seems defenseless against oversimplified, demagogic solutions offered by shortsighted, irresponsible politicians. The understanding of the structure and function of the society has never been as important as today for governance and this task has never been so complex.

The new challenges require an entirely new approach, that of complexity science. Fortunately, the development of information and communication technology is helpful in this endeavor: It produces a flood of data, representing traces of almost all kinds of activities of individuals, thus enabling an entirely new form for social analysis. The future of complexity science will be certainly data driven.

» Results can be used for better or worse «

There is no single discipline which can cope with the complex tasks of our times. That is why social scientists, physicists, mathematicians, biologists, computer scientists, and economists have realized the need of joining efforts. This will become even stronger: The future of complexity science will be increasingly multi-disciplinary.

The accumulation of severe complex problems is expected to motivate decision makers to open their ears to the methods and solutions of complexity science. Network science, the analysis of the massive information and communication data, and large-scale multi-agent modeling with participatory aspects have made this new discipline an applicable tool to handle issues of major concern. The future of complexity science will be increasingly application-oriented.

We are at the dawn of a new era in science. As it is always the case, results can be used for better or worse. It is the responsibility of scientists and decision makers to place complexity science at the disposal of the well-being of people.

János Kertész / Central European University, Budapest
received his Dr. rer. nat. from Eötvös University and obtained fellowships from the Hungarian Academy of Sciences, DAAD, and the Humboldt Foundation. He was postdoc at the University of Cologne and TU Munich, and researcher at the Institute of Technical Physics in Budapest. Since 1991 he has been professor at the Budapest University of Technology and Economics (now part time), since 2012 at the Center for Network Science of the Central European University. He is elected member of the Hungarian Academy of Sciences. His main interest is in interdisciplinary applications of statistical physics.

For a system to be more than the *sum of its parts*, it must have strong, long-range correlations, otherwise it could be divided into essentially independent subsystems. In this sense, complex systems are nonlocal. If the constituent parts of such a system are not uniform, the system itself will depend on a high number of details, it will be irreducible, its description will need a large number of parameters, where even tiny details may matter and fundamentally alter the behavior of the system. Such systems cannot be described by a limited set of explicatory variables, they are intrinsically high dimensional.

This situation is very different from the traditional approach in science where one strives to identify a few important variables and disregard the rest. The old strategy led to spectacular progress, especially in physics, and was consequently emulated also in fields such as economics, where the intrinsic complexity of the subject has doomed it to failure. The full acceptance of the complexity of the environment, society, or economy demands novel research methodologies.

Decision makers are particularly prone to putting their faith into oversimplified models or economic and social theories, often into myths and prejudice, and act under the delusions created by these. In response to a particular problem (unemployment, inflation/deflation, growth, immigration, etc) they tend to introduce measures, regulation, and incentives that act on that particular dimension, thereby inducing a host of collateral changes in a number of other social and economic dimensions, some of them not even visible until later, that combine in a totally unforeseeable manner into an impact on the original variable, which is often paradoxical, counterintuitive, and contrary to the effect the original measure was supposed to achieve. This phenomenon is the

» THE LAW OF UNINTENDED CONSEQUENCES «

content of the law of unintended consequences. Recent examples include large scale biofuel production, international financial regulation, and the monetary policy of low interest rates. By a thorough analysis of such examples, complexity science must shake the belief in oversimplified economic and social theories, must spread its findings over the scientific, business, and policy making communities, and must alert the public to the futility and dangers of voluntarism, blind or misguided decision making, and the false promises of lunatic populism. Via a wide interdisciplinary collaboration, complexity science must learn to build realistic, high dimensional models that allow testing the expected consequences of planned measures, to continuously monitoring the performance of the model during the real-world implementation, and to fine tuning the parameters as the consequences become visible. On the long run, this has to lead to a completely new political culture.

Some priority areas are the environment, water and food scarcity, epidemiology, critical infrastructures, demographic and migration issues, the international order, and security challenges.

Decision makers are acting on complex systems, their natural language must be complexity science. The Complexity Science Hub Vienna will have a large field where it can contribute to the solution of problems that threaten the survival of civilization.

Imre Kondor / Parmenides Foundation, Pullach b. München
is a retired professor of physics, honorary professor of finance at Corvinus University, Budapest, and professor at the Parmenides Foundation. Before retiring, he was a professor of physics from 1989 to 2011 at Eötvös University, Budapest; 2002–2010 permanent fellow, 2002–2008 rector of Collegium Budapest–Institute of Advanced Study. He obtained his MSc from Eötvös University (1966), the CSc (1984) and DSc (1988) degrees from the Hungarian Academy of Sciences. His research experience includes the theory of condensed Bose systems, critical phenomena, disordered systems and spin glasses, and, presently, the application of statistical physics methods to problems in economics and finance.

I want to begin these proceedings by giving some prominence to an elemental tension in the construction of creative institutions. One origin of tension is described by the pragmatist philosopher Charles Peirce: *I do not call the solitary studies of a single man a science. It is only when a group of men, more or less in intercommunication, are aiding and stimulating one another by their understanding of a particular group of studies ... that I call their life a science.*

The countervailing source of tension is well known – research is conducted by individuals or research collaborators – it is not conducted by themes, projects, topics, mission statements, or programs. Here is Max Planck reflecting on the topic: *New scientific ideas never spring from a communal body, however organized, but rather from the head of an individually inspired researcher who struggles with his problems in lonely thought and unites all his thought on one single point which is his whole world for the moment.*

If Peirce is correct, we need to find extraordinary mechanisms for fostering communication among researchers. If Planck is correct, we need to invest most of our energies into finding and supporting singular minds. In a combined approach individuals do the work, largely on their own or with groups, but hiring and advisory committees determine the foci of inquiry – partly through initial forms of support and then on through ongoing, differential funding.

» WE IGNORE THE DISCIPLINES «

This is not how the Santa Fe Institute has proceeded. The way in which SFI has sought to encompass theme and individual is to proceed along at least four different paths that represent efforts at overcoming the institutional polarities: i) Support a community of exceptional individual researchers, with a proven record of collaboration, with the stated ambition to search for fundamental unifying principles in the evolved world – the search for unity is the mechanism of thematic integration. ii) We ignore the disciplines – departments and schools, and visiting committees, and articulate objectives in terms of complexity science – not to be confused with *anything goes science* – and do this with a distributed network of committed researchers. iii) Build a community and a physical institution that is to some extent irresistible to smart, creative individuals, and to build it upon the restless philosophical foundations and iconoclasm of the American frontier. iv) Make this research work available, of interest, and of value to the culture at large – from educators to businesses and policy makers – without becoming a degree-granting institution and without becoming a think tank.

The particular work that gets done, whether the statistical physics of the economy, the nature of intelligent systems, the dynamics of evolving networks, or the evolution of the health system, will be determined through harnessed serendipity, what Francis Crick and Murray Gell-Mann refer to with admiration as frozen accidents.

David Krakauer / Santa Fe Institute
is president and William H. Miller Professor of Complex Systems at the Santa Fe Institute. David's research explores the evolution of intelligence on earth. David has been a visiting fellow at the Genomics Frontiers Institute at the University of Pennsylvania, a Sage Fellow at the Sage Center for the Study of the Mind at the University of Santa Barbara, a long-term fellow of the Institute for Advanced Study in Princeton, and visiting professor of evolution at Princeton University.

The last few years have seen the emergence of the *sharing economy*. As social media blurred the distinction between author and reader, everyone can now offer or receive services thanks to the networking tools provided by new technological companies. Take Uber, and its billion of journeys in 2015 alone, with tens of thousands of vehicles crawling every moment in the globe's biggest cities. As often, when confronted with a technological change, we observe a polarization of society, and the search for an equilibrium characterized by new norms, rights, and obligations. Understanding the mechanisms behind this re-organization requires an integrated, interdisciplinary approach, covering an intricate web of legal, societal, economical, and computational issues which, we believe, could benefit from a complex systems perspective. As a first step, we are currently studying the dynamics of pricing in Uber. In this new de-regulated world, journey prices fluctuate in time depending on traffic but also on the service's perceived balance of passenger demand and driver supply.

» Social media blurred the distinction between author and reader «

Even worse, the algorithmic rules driving surge pricing are known by the operator alone. In order to bring back knowledge and transparency in urban transport, we have been developing OpenStreetCab. A collaboration amongst three European Universities (Cambridge, Lancaster, and Namur), the project aims to provide passengers with knowledge on pricing for their taxi journeys in real time. The concept is simple. Users submit a query with information on their intended trip. The app responds with estimates on the costs incurred by major providers for their journey. In the background, the service is powered by an integration of data mining and mobility models, in order to predict journey prices, and real-world experiments and interviews in order to validate the predictions. Uber may have disappeared in 10 years' time, just like OpenstreetCab, but my hope is that similar initiatives will emerge in future years in order to help towards our transition into a world where technology is dominant and at the same time remains competitive and fair.

Renaud Lambiotte / University of Namur
is professor in the Department of Mathematics at the University of Namur, and director of the Namur Center for Complex Systems. He received his PhD in theoretical physics from Université libre de Bruxelles in 2004, and has been a Research Associate at ENS Lyon, Université de Liège, Université catholique de Louvain and Imperial College London. His research interests include network science, data mining, stochastic processes, social dynamics, and neuroimaging. He has published over 70 peer-reviewed articles, including two PNAS and two Nature Communications, with a total of around 8000 citations. He is also the co-founder of L'Arbre de Diane, a publishing company at the interface between science and literature.

Compared to the physical and biological sciences, so far complexity has had far less impact on mainstream social science. This is not surprising, but it is alarming because we find ourselves in the midst of a planetary-scale transition from the Holocene to the Anthropocene. We have already breached some planetary boundaries for sustainability, but those tipping points are nearly invisible from the perspective of the linear equilibrium models that continue to hold sway in social science. For example, in 2014 the G20 nations agreed to seek between US$ 60 trillion and US$ 70 trillion in new infrastructure investments by 2030, which would more

» INFRASTRUCTURE TSUNAMI COULD EASILY DWARF CLIMATE CHANGE «

than double the global total value of infrastructure. Roughly 90% of the new projects are in developing nations, often in the tropics or subtropics which harbor the planet's biologically richest and environmentally most critical ecosystems. The environmental impacts of this infrastructure tsunami could easily dwarf climate change and the acidification of the oceans, as thousands of projects penetrate into the world's last surviving wild areas. The dynamics of coupled human-environmental systems needs to move from the periphery to the forefront of research and teaching in the social sciences.

Stephen Lansing / Nanyang Technological University, Singapore; Santa Fe Institute

co-directs the Complexity Institute, as well as is the Faculty Associate Chair for Asian School of the Environment at Nanyang Technological University in Singapore. He is also an external professor at the Santa Fe Institute, an emeritus professor of anthropology at the University of Arizona, and a senior research fellow at the Stockholm Resilience Centre. Before moving to Arizona in 1998, Lansing held joint appointments at the University of Michigan in the School of Natural Resources & Environment and the Department of Anthropology, and earlier chaired the anthropology department of the University of Southern California.

Complexity is a highly interdisciplinary science. Although there are drawbacks for researchers to work at the interface of different fields, such as the cost to set up common languages, and the risks associated with not being recognized by any of the well-established scientific communities, some of my recent work indicates that interdisciplinarity can be extremely rewarding. Drawing on large data sets on scientific production during several decades, we have shown that highly interdisciplinary scholars can outperform specialized ones, and that scientists can enhance their performance by seeking collaborators with expertise in various fields. My vision for complexity is based on the added value of its interdisciplinary nature. I list below three research directions that I am personally eager to explore, and that I think will be among the main challenges of complexity in the next 10 years.

i) Big networks in the service of humanity. Big data and networks can help mankind to tackle big problems. For instance, DataKind is an organization launched in the United Kingdom in 2013 to use data for good. Projects range from the use of networks to help charities maximize their impact in society to the development of data-driven decisions to create sustainable cities. As another example, in a collaboration between my research group and the Startup-Network Lab, we are studying the worldwide network of social/professional interactions between start-ups to characterize the success of certain innovation ecosystems and to help policy-makers devise effective data-informed decisions to nurture the growth of certain areas. The potentials of complexity scientists to address critical societal issues in the fields of health, education, poverty, the environment, and cities are incredibly high.

ii) Functional brain networks. The human brain is the most complex of complex systems. Different brain imaging techniques allow looking at the correlations between the activities of the different regions of our brain in the form of a network (functional brain

» COST TO SET UP COMMON LANGUAGES «

network). Some of these techniques are becoming highly portable, so that wireless brain-scanning helmets collecting real-time functional networks and delivering them directly to a mobile device are now available. The opportunities to investigate the complex functioning of our brain while we are performing a physical or an intellectual task are practical infinite (PLoS ONE 5(12):e14187, 2010). New-generation data to study will consist of recordings of functional brain networks while we enjoy a painting or create an artistic work, or when we walk in the most beautiful or in the less pleasant streets of our city.

iii) The *complexity manager*. Tech and non-tech business companies are increasingly hiring data scientists. As a scientific community we can be considered successful only if we will be able to produce a well-defined new professional role, that of a *complexity expert* or *complexity manager*. Although I am not completely sure of the name, I am sure of the important role that a transdisciplinary approach can play in any business, and it would be nice to see complexity managers in every company or organization in the near future. We have to keep this in mind in the education of next-generation complexity scientists, so that researchers working in complexity will be able to pursue insights without sacrificing their career progress with respect to ultra-specialized scientists.

I am sure the Complexity Science Hub Vienna will contribute in each of these work directions.

Vito Latora / Queen Mary University of London
is chair of complex systems and head of the Complex Systems and Networks Group at the School of Mathematical Sciences of Queen Mary University of London. Vito studies the structure and the dynamics of complex systems using his background as theoretical physicist and methods of statistical physics and nonlinear dynamics to look into biological problems, model social systems, and design complex networks. He is currently interested in the mathematics of multiplex networks, and is working with neuroscientists and urban designers to understand the growth of networks as diverse as the human brain or the infrastructures of a city.

In his famous historical account about the origins of molecular biology Gunther Stent introduced a three phase sequence that turns out to be characteristic for many newly emerging paradigms within science. New ideas, according to Stent, follow a sequence of romantic, dogmatic, and academic phases. One can easily see that complex systems science followed this path. The question now is whether we are in an extended academic phase of gradually expanding both theoretical and practical knowledge, or whether we are entering a new transformation of complex systems science that might well bring about a new romantic phase. I would argue that complexity science, indeed, is at the dawn of a new period – let's call it complexity 3.0. The last academic phase has seen the application of complex systems ideas and methods in a variety of different domains. It has been to a large extent business as usual.

»ADDRESS THE MAJOR SOCIETAL CHALLENGES«

What we need today and in the future is a new version of complex systems science that deepens our theoretical understanding and helps us address the major societal challenges humanity is facing right now. Questions in need of an answer include but are not limited to the nature of historical and evolutionary processes and their effects on both enabling and constraining the future; the nature of cognition at all levels – from simple systems to brains, collectives, and cities; and the nature and future of intelligence and life, both natural and created, including AI and AL. This last point brings us to the bridge between theoretical and practical knowledge. We are not just detached observers but active creators of complex systems and the history of our past interventions and creations has produced the challenges and uncertainties of the present and future. For us to have any chance of orderly survival, we need to apply the best possible insights into complex systems to the challenges societies are facing today. The challenges are huge, but the energy of a new romantic period also gives us hope.

Manfred Laubichler / Arizona State University; Santa Fe Institute
is President's Professor of Theoretical Biology and History of Biology at Arizona State University. At ASU he serves as director of the ASU-Santa Fe Institute Center for Biosocial Complex Systems and associate director of the Origins Project. Besides his appointments at ASU, Laubichler is an external professor at the Santa Fe Institute and at the KLI, an institute for advanced study in natural complex systems in Klosterneuburg, Austria; a visiting scholar at the Max Planck Institute for the History of Science in Berlin, Germany; and an adjunct scientist at the Marine Biological Laboratory in Woods Hole, MA.

Our societies are being thoroughly transformed by the pervasive role technology is playing in our culture and everyday life. Nowadays the term *techno-social systems* is adopted to quickly refer to social systems in which the technology entangles, in an original and unpredictable way, cognitive, behavioral, and social aspects of human beings. This revolution does not come without a cost and in our complex world new global challenges always emerge that call for new paradigms and original thinking: climate change, global financial crises, global pandemics, growth of cities, urbanization, and migration patterns. In this framework we progressively face the need to increase the number of people able to imagine original and valuable solutions to sustain large human societies safely and prosperously.

Creativity is a key factor in the evolution of human societies since they represent the primary motor to explore new solutions in ever-changing and unpredictable environments. New technological artifacts, scientific discoveries, and new social and cultural structures are very often triggered by mutated external conditions. Unfortunately, the detailed mechanisms and the forces through which humans and societies express their creativity and innovate are largely unknown. No comprehensive conceptual, computational, or mathematical frameworks have been proposed so far to understand the structure of the space of the *possible*, how it gets re-structured, and how people explore it. As a consequence our societies and their components constantly struggle to detect trends and foresee emerging opportunities. Still, creativity and innovation have proved to be resistant to measurement, understanding, and control.

Nowadays, complexity science has for the first time the possibility to exploit the concurrence of three extraordinary circumstances: i) the possibility we have now to perform

» WHERE CREATIVITY AND INNOVATION ARE MUCH NEEDED FUELS «

a tomography of our societies, with a key contribution of people acting as data gathering *sensors*; ii) the opportunities web-gaming and social computation are offering to the emergence of new forms of participation arising from the interplay of ICT services and communities of citizens; iii) the maturity of complex systems and data science applied to socio-technical systems.

The concurrence of all these elements is opening tremendous opportunities towards an understanding of the complexity of our societies, with the final goal of deploying human imagination for the betterment of our communities and even civilization. From this perspective the Complexity Science Hub Vienna is promising to play a leading role in this endeavor that blends in a unitary interdisciplinary effort three main activities: data science, theoretical modeling, and web-based experiments. Unveiling and quantifying the complex ecology of creativity and innovation has the potential to impact sectors – education, learning, research, social challenges – where creativity and innovation are much needed fuels.

Vittorio Loreto / Sapienza University, Rome; ISI Foundation
is professor of physics of complex systems at Sapienza University of Rome and research leader at the ISI Foundation. Loreto is coordinating the Co-laboratory of Social Dynamics spread among Sapienza University and the ISI Foundation. His scientific activity is mainly focused on the statistical physics of complex systems and its interdisciplinary applications. In the last few years he has been active in the fields of granular media, complexity and information theory, complex networks theory, communication and language evolution and social dynamics. He coordinated several projects at the EU level and has recently coordinated the EU project EveryAware. He is presently coordinating the KREYON project aimed at unfolding the dynamics of innovation and creativity. He published over 120 papers in internationally refereed journals and chaired several workshops and conferences.

Society currently generates a gargantuan amount of new data each day and a significant amount of these data can be described and modeled in terms of networks and/or flows in them. One ubiquitous character of complex systems is the heterogeneity of their components, of their relationships, and of their pair similarities. To go beyond the detection and modeling of heterogeneity, it is highly informative to filter out features and relationships that cannot be explained by a random null hypothesis taking into account the heterogeneity of the system. Information filtering performed on networks and, more generally, on complex systems allows researchers to detect and characterize structures and phenomena that are present in the system of interest. Statistically validated networks, i. e. networks obtained by performing a statistically controlled process of information filtering, select

» THE APPROACH OF STATISTICALLY VALIDATED NETWORKS ALSO DETECTS UNDER-EXPRESSIONS «

over-expressed relationships that are highly informative on the mesoscopic structure of networks where clusters (communities) are frequently detected. The approach of statistically validated networks also detects under-expressions of relationships that have a statistical meaning and can contribute to quantify aspects, e. g. negative relationships that are usually not reported or difficult to obtain in the construction of networked structures. A few methods producing statistically validated networks are already available for binary or weighted networks. During the coming years information filtering methods applied to networks and complex systems should be expanded, new methods should be proposed, and their role in network characterization and network modeling clarified and formalized.

Rosario Nunzio Mantegna / Central European University, Budapest
is professor at Central European University Budapest, at Palermo University, and honorary professor at University College London. He was postdoc at the MPI for Quantum Optics in Munich and at Boston University. His research covers interdisciplinary applications of statistical physics. He is one of the pioneers in the fields of econophysics and economic networks. Rosario has been principal investigator or member of several international and national research projects.

In the last 20 years or so, the field of complexity science has entered a new age. The combination of new theoretical insights and the data revolution has prepared the ground for a number of conceptual milestones in many disciplines as diverse as biology, physics, engineering, and economic and social sciences. At the same time, we have been able to identify new challenges whose solutions will confer the science of complex systems an unprecedented applied dimension. Here I would like to focus on one of these challenges: the socio-technical man. With the ever-increasing growth of both the world population and new technologies, it is fundamental for the well-being of humanity and our society to understand how humans interact among them and with the new technological environment.

Can we anticipate social collective phenomena and human behavioral responses? If so, under which scenarios and at which scales? Do we have data to answer the aforementioned challenge? Will big data be sufficient for that purpose or is there a need for more controlled, on-demand experiments? Is our online behavior the same as our offline reactions or responses? Or, maybe the biggest question of all: Do we know at all the laws governing human behavior? Perhaps these key questions constitute some of the major challenges in complexity science. Socio-technical systems exhibit emerging dynamics and are often out of equilibrium. They also involve many temporal and spatial scales, are governed by nonlinear effects, and can often adapt to both external and internal perturbations. That is, we have roughly all the ingredients of a truly complex system!

» THE SOCIO-TECHNICAL MAN «

If we aim at successfully describing the socio-technical man, we need to understand, among others, some basic problems such as how humans interact with the environment and other humans, how cooperative behavior emerges and survives, and how social networks evolve and shape the way we communicate and interact with each other.

To this end, we should develop new ways to analyze existing data, perform controlled experiments with groups – of different sizes – of humans facing social dilemmas and hypothetical scenarios and come out with new theoretical and computational methods and algorithms. By doing this, we will be in the position to better understand the factors determining human behavior in a plethora of situations, and hopefully provide hints to, e. g., policy makers with the aim of creating a better and more sustainable society for the future. On its turn, given the universality of many concepts and methods in complexity science, we are sure that the new methods will also contribute to the development of other areas (for instance, to the physics of non-equilibrium systems, economics, and even ecological problems).

Finally, I would like to mention that to achieve the aforementioned goals, it is mandatory to work in multi- and interdisciplinary teams, and in this sense, the Complexity Science Hub Vienna is a unique opportunity to tackle some of these challenges.

Yamir Moreno / Cosnet Lab; University of Zaragoza
received his PhD in physics (Summa Cum Laude, 2000) from the University of Zaragoza. Shortly after, he joined the Condensed Matter Section of the International Centre for Theoretical Physics (ICTP) in Italy as a research fellow. He has been the head of the Complex Systems and Networks Lab (COSNET) since 2003 and is also affiliated to the Department of Theoretical Physics of the Faculty of Sciences, University of Zaragoza. During the last years he has been working on multilevel complex systems and the structure and dynamics of networks. Prof. Moreno's scientific production amounts to 155+ peer-reviewed publications with a total of 9700+ citations.

My vision of complexity sciences targets their potential to extend the range, precision, and depth in making predictions. While this has always been the ambition and yardstick for the physical-mathematical sciences, complexity sciences now allow to include society and social behavior – to some extent. There is agreement that society is a complex adaptive system, CAS, with a few peculiarities. Ignoring, downplaying, or *naturalizing* them, i.e. to take them as essential and given, carries the risk to end up with abstractions which are cut-off from the dynamics of societal contexts. One of the peculiarities of society as a CAS is that the models with which we try to make sense of the world are invented and constructed by us. It is humans who make observations and provide the assumptions on which models are based. Humans leave traces that are collected and processed to be transformed into data. Humans decide to which purpose they will be put and how they will be repurposed. Humans are object of research and subject. Coping with these peculiarities requires an in-built reflexivity. Practioners must perform a double act and do so repeatedly. They must engage in a focused way with their scientific work and equally distance themselves by critically reflecting their often tacit assumptions. A friend of mine, Yehuda Elkana, called this two-tier thinking.

The other peculiarity of society as a CAS is that it is replete with the unintended consequences of human action. Evolution has equipped us to recognize immediate cause-effect relations. Cultural evolution, spearheaded by the amazing

»THE EMBARASSMENT OF COMPLEXITY«

achievements of science and technology and their transformative power on societies, turns out to be much faster than biological evolution. Complexity emerges from manifold interactions. The denser human networks are and the greater the number of interactions, the more unintended consequences will be generated. We do very poorly in anticipating them, let alone have a reasonable grasp in understanding them. Systemic societal risks are also the result from the unintended consequences of human action. They are largely invisible and prone to eruption. We have to move closer to discover the fault-lines, even if, like with earthquakes, to predict the timing is still far away.

At a time which some see as a break-down of order and many anxieties abound, the embarrassment of complexity is widely felt. It results from the gap between feeling overwhelmed by the sheer scale and scope of problems confronting us and the pressure to pretend that we can manage them. In such messy encounters clinging to old certainties will not get us very far. We should also beware of new certainties, often offered by those who thrive on inducing fear. Therefore, my message – and vision – for the Complexity Science Hub Vienna is to encourage you to embrace uncertainty. It acknowledges the radical openness of the future, while offering the key to deal rationally with it.

Helga Nowotny / ETH Zurich
is professor emerita of social studies of science, ETH Zurich, and a founding member of the European Research Council. From March 2010 until December 2013 she served as ERC president. Currently she is a member of the Austrian Council and vice-president of the Council for the Lindau Nobel Laureate Meetings. Helga Nowotny holds a PhD in sociology from Columbia University, NY, and a doctorate in jurisprudence from the University of Vienna. She has held various teaching and research positions, among others at King's College, Cambridge, and École des Hautes Études en Sciences Sociales, Paris. Helga Nowotny is foreign member of the Royal Swedish Academy of Sciences.

My vision for complexity is in the first place visual and has the honor of covering this book. I tried to find an image that represents both the problem and the potential for a solution.

Why do I paint on maps? I generally like to ask if our view of the world can be different. Seeing it from the same perspective for long usually feels dangerous to me, because we can only have an overview if we keep moving. Humans give a lot of structure to the world, which is ok – but how can we have our structures flexible enough for constantly changing circumstances in a worldwide dimension? How can we explore what is possible while keeping an eye on what is necessary? The vertical mathematical view of our maps is often useful but also abstract, and the political frontiers of today cause a lot of problems. With my image I try to visualize my hope that humans are finally capable of handling complex issues and find forms of respectfully living together. The horizon is an imagined limit and reminds me of the fact that, as individuals, we can only see a small part of the possibilities that might be in our reach – seen from above those limits do not exist. But as an individual we are not above, so we better share our perspectives, investigate the unknown, and move towards it together.

But arts and sciences should be like mines, where the noise of new works and further advances is heard on every side.
SIR FRANCIS BACON

Art can – in whatever form – express an observation, involve people, make them think, and argue on issues like complexity. Science analyses certain phenomena and formulates methods for advancing concrete problems, which gives me hope. Art is able to help us accept that there will probably always be

» FORGET ABOUT OUR CONCEPTS AND START PLAYING «

phenomena that can't be fully explained. It asks questions, and through its emotional qualities it can also make us stand back from a problem without having an answer to it right away. Art tries to put order into complexity rather by description than by explanation. Its forms of prediction are less concrete than scientific ones. No artist could predict the economic crash in 2008, but many of them – like Damien Hirst with his *Golden Calf* – had already pointed at the madness of financial interests. No artist did *know* that World War II would come, but George Grosz surely did not feel comfortable with the politics during the Weimar Republic when he painted and exhibited his *Pillars of Society* in 1926. A poetic warning is also a vision for complexity.

Why should a seemingly *irrational* approach not open the door for a *rational* solution? When I once had to draw an image for a contemporary theatre poster I finally found the inspiration in an encyclopedia of night butterflies. Provoking chance by spinning around in areas we haven't yet been to can apparently be useful. Trying to forget our concepts for a moment and start playing is a popular method in visual arts.

If music theorist Brian Eno is right by assuming that artists rather intuitively seek for alternatives while scientists want to find out how things actually work: let's give each other a hand as often as possible – like in this book!

Olaf Osten / artist and graphic designer based in Vienna
was in charge of this book's visual appearance. His work has been presented in a range of exhibitions in several countries and is represented eg in the collections of the International Peace Institute, the Vienna Museum and the Austrian Chamber of Labor. As a graphic artist he is active in a variety of interdisciplinary contexts, with co-operations including the Wiener Festwochen Theatre Festival, Museum of Modern Art Vienna, and Impulstanz International Dance Festival. In 2012 he received together with artist Ernst Logar the State Prize for Austria's Most Beautiful Books.

In our societies, we invest a lot of resources to ensure technological progress, and to find out secrets about the deepest corners of our universe. As a result, we are experiencing technological breakthroughs at an unprecedented rate, and we invest billions into research that is as far away from everyday life as the east is from the west. While technology has the potential to improve the quality of life and the satisfaction of curiosity is rewarding, we must not fail to notice that, at the same time, many of our societies are falling apart. Inequality all over the world has become staggering, and we are seriously failing to meet the most basic needs of the majority of people that live on this planet. We have neglected adverse societal changes for far too long, and we have done very little in terms of research to understand, and more importantly, to reverse, or at least to impede, these very worrying trends.

We need to rethink and learn to cooperate, and we have to understand that our actions have consequences that go far beyond our local communities. But to cooperate is in a way unnatural. If only the fittest survive, why should one perform an altruistic act that is costly to perform but benefits others? Why should we care for and contribute to the public good if free-riders can enjoy the

» Pave the way towards better human societies «

same benefits for free? Existing research indicates that a comprehensive answer to these questions requires that we look beyond the individual and focus on the collective behavior that emerges as a result of the interactions among individuals, groups, and societies. Cooperation in human societies is an emergent, collective phenomenon in a complex system. An important vision for complexity, for complex systems science, and for the Complexity Science Hub Vienna is therefore to be at the forefront of research that will pave the way towards better human societies.

The aim should be to merge the most recent advances in the social sciences and methods of statistical physics and network science for modeling and describing the rise and fall of cooperation in human societies. We must develop a predictive, computational theory that will allow us to better understand the rich variety of phenomena that rely on large-scale cooperative efforts. From the mitigation of social crisis and inequality to the preservation of natural resources for next generations, by having a firm theoretical grip on human cooperation we can hope to engineer better social systems and develop more efficient policies for a sustainable, better future.

Matjaž Perc / Complex Systems Center Maribor; University of Maribor
is professor of Physics at the University of Maribor and director of the Complex Systems Center Maribor. He is member of Academia Europaea and the European Academy of Sciences and Arts, and he was among top 1% most cited physicists according to Thomson Reuters Highly Cited Researchers in 2014 and 2015. He received the Zois Certificate of Recognition for outstanding research achievements in theoretical physics in 2009 and the Young Scientist Award for Socio-and Econophysics in 2015. His research on complex systems covers evolutionary game theory, social physics, large-scale data analysis, and network science, and has been covered by Nature News, New Scientist, MIT Technology Review, Inside Science, Boston Globe, Fox News, Chemistry World, Phys.org, Physics Today, Science News, and APS Physics.

The immense accumulation of data is a new phenomenon which induces many considerations, represents great potential, and sometimes leads to mythical expectations. Here we discuss a specific example of big data applications, the case of economic complexity. This is a new perspective on fundamental economics, adopting a bottom-up approach which starts with a novel use of older data and then develops into its own streamline. The approach confirms some expectations about big data but also disproves others.

Take home message 1: As soon as you have a new idea and a new algorithm you immediately realize that the data available (originally collected for different purposes) is not

» IT SHOULD BE STUDIED AND TAILORED TO EACH PROBLEM «

optimal and you want more data of a new type. There is no infinite dataset one may collect a priori which is good for all problems. Big data is also big noise. Data selection is crucial to optimize the signal to noise ratio.

Take home message 2: One is often led to a complex network but this does not necessarily imply the use of Google page rank. For fundamental economics we argue that a different algorithm (non linear) is more appropriate. This suggests that for each problem there should be a clear understanding of what the relevant information is and how to extract it from the data. This cannot be a single recipe for all fields of analysis. Instead, it should be studied and tailored to each problem.

Luciano Pietronero / Sapienza University, Rome
studied physics in Rome and was a research scientist at Xerox Research in Webster (1974) and Brown Boveri Research Center (CH) 1975–1983. He then moved to the University of Groningen (NL), where he was professor of condensed matter theory (1983–87). Since 1987 he is professor of physics at Sapienza University Rome. He is also founder and director of the Institute for Complex Systems of CNR (2004–2014) and received the Fermi Prize in 2008. Research interests include condensed matter theory, statistical physics, complex systems and economic complexity.

Complexity science is for sure a challenging interdisciplinary field of research which will provide new useful perspectives and novel tools for a more efficient participatory society in the next years. Groundbreaking discoveries and innovation are the results of the interplay of many different factors which emerge in a non-linear and often unpredictable complex way from the fruitful bottom-up mixing of different disciplines. This fact, however, although advocated by many, is unfortunately very rarely fostered and put in practice. Nowadays, funding agencies, often forgetting that research means working at the edge of human knowledge and can be also unsuccessful, rarely tend to risk their money in unconventional proposals regarding complexity. They are more inclined to support scientists, projects, and ideas that have a well settled successful past and operate within well established and more conventional fields of research. This is certainly not a good policy for promoting innovative results. As stigmatized in a recent editorial by Nature (Take more risk, 528, 8/2015), such policies stimulate a conservative and not very efficient behavior, which discourage young scientists and force them to a scientific conformism. The same happens for careers. Today it is not easy for a scientist in complex systems to get a permanent job. Agencies

» FUNDING AGENCIES RARELY TEND TO RISK THEIR MONEY «

adopt risk-averse strategies forcing publishing along mainstream research directions. In this way it is very difficult to attract young scientists towards multidisciplinary research. Today it is more convenient to work along well known consolidated paradigms. On the other hand, working in complexity often implies a less orthodox vision of science, not having fear of hybridize your own education with that of others, stimulate bottom-up novel ideas, encourage young scientists to think outside the box and educate them to listen to other disciplines to create novel ways of looking at our complex interconnected world. We need all this to solve the many intriguing problems affecting our present way of living, to find new sustainable ways to exploit our limited resources, to discover innovative ways to conceive our mobility, our communication systems, our financial markets, and to live better in a more democratic and participatory society.

It is for these reasons that a new complexity hub in Vienna is certainly very welcome and will play a fundamental role for the development of this field. It will attract scientists with different backgrounds from all over the world, giving them the possibility to work successfully together in a stimulating and unique atmosphere.

Andrea Rapisarda / University of Catania
is professor of theoretical physics at the University of Catania, Italy. He is also coordinator of a PhD course in Complex systems for physical, socio-economic and life sciences. *He is co-author of more than 120 publications in international journals and member of the editorial board of Physica A, Heliyon, Cogent Physics, International Journal of Statistical Mechanics. His main interests of research are statistical mechanics, complex networks, and multi-agent models applied to socio-economic systems.*

Cybernetics, systems science, synergetics, global systems, complexity - what is new in understanding how the whole works? There are two new opportunities that pose at least two challenges: the availability of massive data and the direct social relevance of the field of research.

Data: What do we understand when we know everything? The challenge is to understand mechanisms that would allow understanding data. It is not only to infer behavior from statistical correlations, but to establish cause-effect relations from isolated mechanisms. This requires defining clear questions that can be answered from simple models: Simplicity is the ultimate sophistication (Leonardo da Vinci).

» WHAT IS NEW IN UNDERSTANDING HOW THE WHOLE WORKS «

Social relevance: The challenge is to convey two messages to the society at large: i) Complexity science deals with transversal problems, such as multiscale problems, which are deep, fundamental science at the frontiers of knowledge. ii) Complexity science provides perspectives and concepts that are absolutely needed in our daily life and that defy common wisdom. For example, individual intentions cannot be inferred from observation of social aggregates, or a median income is a meaningless concept, as well as the average number of acquaintances of an individual in a social network.

Maxi San Miguel / IFISC (CSIC-UIB), Palma de Mallorca
is professor of physics at the University of the Balearic Islands and director of IFISC (Institute for Cross-Disciplinary Physics and Complex Systems), Palma de Mallorca, Spain. His academic career includes positions at the University of Barcelona, Temple University (Philadelphia), Sapienza University Rome, University of Arizona, University of Strathclyde (Glasgow), and Helsinki University of Technology. His research activity spans across different fields of statistical and nonlinear physics (stochastic processes, phase transitions, pattern formation and spatio-temporal complexity, complex networks), computational social science, and laser physics and photonics. He is the author of over 400 papers in top journals of physics, engineering, ecology, social science and multidisciplinary science.

Every week, more than 1 million people are currently being added to cities across the globe. This unprecedented trend of urbanization, together with growing concerns over energy supply and climate change, rapidly outpaces existing approaches for the planning and design of cities. A prominent warning example is Beijing's recent failure to implement a multi-centered urban form that has led to counter-intuitive people flows, immense traffic congestion, and air pollution. Thus, a new quantitative understanding of cities is urgently needed to reduce the risks of such detrimental planning outcomes and to eventually build more sustainable and more livable urban spaces.

Complexity science is particularly promising: it offers the appropriate *big-picture* view and mathematical tools to elucidate fundamental regularities in the dynamics of cities. Perhaps most importantly, it allows us to better anticipate systemic behavior that results from the many interactions of all the components that make up a city (including people, infrastructures, etc.). The recent discovery of *urban scaling laws* for the prediction of socio-economic outcomes and their explanation through the network of social interactions underpin the potential of complexity science towards an integrative understanding of cities.

At the same time, we are witnessing an ever-growing availability of large-scale data on human activities, ranging from mobile phone records to GPS traces and location-based social

» IT OFFERS THE APPROPRIATE ›BIG-PICTURE‹ «

networks. Hence, for the coming decade, I see it as the next logical step to tackle the big problems of cities by enriching complexity science with big data analytics. The combination of these two burgeoning fields will enable us to develop simple mathematical models for the dynamics of cities based on highly fine-grained spatial and temporal considerations. This new quantitative understanding of how people actually make use of urban space and what the consequences for the overall city performance are, is ultimately key to a high quality of urban life at significantly reduced environmental and social impacts.

The Complexity Science Hub Vienna will be a unique catalyst for the advancement of our understanding of complex systems by making use of big data. It will bring together people with highly diverse backgrounds, ideas, and scientific toolkits. As such, the Hub will provide the perfect intellectual environment to develop the interdisciplinary research that is needed to tackle the immense challenges of urbanization. Last but not least, Vienna is an inspirational setting for the study of cities, being regularly named the world's top city to live in.

Markus Schläpfer / ETH Future Cities Laboratory Singapore
is a principal investigator at the ETH Future Cities Laboratory Singapore, where he has been leading the Urban Complexity project since 2016. He was a postdoctoral fellow at the Santa Fe Institute (USA) and at MIT's Senseable City Lab (USA). He received his PhD in 2010 from ETH Zurich (Switzerland). His main research goals are the derivation of quantitative, predictive models for the organization of cities and its interplay with urban infrastructure networks. He grounds his research on the increasing availability of large-scale data on human activities and applies methods from complexity science to gain a comprehensive view of the urban dynamics. His work has been featured worldwide, including the New York Times, Nature, The Atlantic, Quartz, MIT Technology Review, and Spiegel Online.

Complexity science can help to understand the functioning and the interaction of the components of a city. In 1965, Christopher Alexander gave in his book *A city is not a tree* a description of the complex nature of urban organization. At this time, neither high-speed computers nor urban big data existed. Today, Luis Bettencourt et al. use complexity science to analyze data for countries, regions, or cities. The results can be used globally in other cities. Objectives of complexity science with regard to future cities are the observation and identification of tendencies and regularities in behavioral patterns, and to find correlations between them and spatial configurations. Complex urban systems cannot be understood in total yet. But research focuses on describing the system by finding some simple, preferably general and emerging patterns and rules that can be used for urban planning. It is important that the influencing factors are not just geo-spatial patterns but also consider variables which are important for the design quality. Complexity science is a way to solve the dilemma of oversimplification of insights from existing cities and their

» SOLVE THE DILEMMA OF OVER-SIMPLIFICATION «

applications to new cities. An example: The effects of streets, public places and city structures on citizens and their behavior depend on how they are perceived. To describe this perception, it is not sufficient to consider only particular characteristics of the urban environment. Different aspects play a role and influence each other. Complexity science could take this fact into consideration and handle the non-linearity of the system.

One purpose of the future urban model is thus the simplification of reality without forgetting that it is embedded in a complex system. By reducing urban phenomena, behaviors, relations and qualities to their essential representing factors, it will be easier to modify and use the urban model for predictions and simulations of future scenarios.

The final goal of urban complexity science must be to increase the sustainability, resilience, responsiveness, and livability of any city in the world. It will develop in the wide-open field between complexity science and design science on the one side and human cognition and design intent on the other side.

Gerhard Schmitt (with Reinhard König, Peter Buš, Johannes Müller, Matthias Standfest), ETH Zürich
is professor of information architecture at ETH Zurich, Lead PI of the ETH Future Cities Responsive Cities Scenario, founding director of the Singapore-ETH Centre in Singapore, and ETH Zurich Senior Vice President for ETH Global. Gerhard Schmitt holds a Dipl-Ing and a Dr-Ing degree of the Technical University of Munich, and a master of architecture from the University of California, Berkeley. His research focuses on urban metabolism with the associated emissions, smart cities and linking big data with urban design, urban models, simulation, and visualization. He and his team developed and taught the first Massive Open Online Courses on future cities and livable cities.

Definitions of complexity are notoriously difficult if not impossible at all. A good working hypothesis might be: Everything is complex that is not simple. This is precisely the way in which we define nonlinear behavior. Things appear complex for different reasons: i) Complexity may result from lack of insight, ii) complexity may result from lack of methods, and (iii) complexity may be inherent to the system. The best known example for i) is celestial mechanics: The highly complex Pythagorean epicycles become obsolete by the introduction of Newton's law of universal gravitation. To give an example for ii), pattern formation and deterministic chaos became not really understandable before extensive computer simulations became possible. Cellular metabolism may serve as an example for iii) and is caused by the enormous complexity of biochemical reaction networks with up

» DEFINITIONS OF COMPLEXITY ARE NOTORIOUSLY DIFFICULT «

to one hundred individual reaction fluxes. Nevertheless, only few fluxes are dominant in the sense that using Pareto optimal values for them provides near optimal values for all the others.

A self-explaining successful combination for bacterial metabolism is maximization of energy production, maximization of biomass production, and minimization of total fluxes (Science 336:601-604). Future understanding of the various complex systems from biology, sociology, economics, and other fields will be achievable only by a concerted approach from all three disciplines: rigorous mathematical analysis and modeling, harvesting enough and the right data, and searching for the best variables in the description of the system. Big data alone without improved theories will not bring the desired progress.

Peter Schuster / University of Vienna
looks back on more than 40 years of a very successful scientific carrier. As one of the trailblazers in the field of molecular evolution, his revolutionary ideas have and still do influence the thinking of many researchers in biology and chemistry. His contributions span from early work in quantum mechanics, over dynamical systems theory and chemical kinetics, to models of the evolutionary process and systems biology. Peter Schuster was long term head of the Institute for Theoretical Chemistry at the University of Vienna and President of the Austrian Academy of Sciences. He was external faculty member of the Santa Fe Institute, and founding director of the Institute of Molecular Biotechnology in Jena, and is a member of the US National Academy of Sciences and the German Academy of Sciences Leopoldina.

My vision is that complexity science will finally become useful. This means we do not only gain groundbreaking insights into the structure and dynamics of complex systems – this goal was already achieved to a large degree in the past three decades. Now, we will turn all these insights, concepts, and methods into something that will improve our socio-technical world – not in a general and abstract manner, but in a detectable and measurable way.

My vision, and my personal research program for the next years, is to achieve a fundamental understanding of resilience in social organizations. Resilience denotes the capacity of a system to withstand shocks and its ability to recover from them. Existing concepts of resilience, e. g. from engineering or population biology, do not lead us very far when it comes to highly volatile organizations as, e. g., collaboration networks of developers producing a specific software. We have massive data about all the interactions between these developers, their entry and exit. But why is it that some projects are able to cope with (internal) crises and (external) shocks, while others simply lose momentum and collapse? And what are the early warning signals of such a possible negative turn?

» Complexity Science will finally become useful «

Big data analytics alone will not tell us the answer if we do not have concepts about what to measure and how to quantify this against available data. Existing knowledge about complex networks provides a solid ground to start from. But it has to be extended to capture the social dimension of interactions. Agent-based modeling will allow us to cope with this. However, improving the response of social organizations to change also requires concepts of how to influence agents. Therefore, there is a need for mechanism design on the level of interactions.

Hence, research on the resilience of social organizations will bring several disciplines together, ranging from social science to applied mathematics. We need to quantify and model social phenomena, beyond simplicity. We are no longer interested in confirming the power law, we are interested in the deviations from the power law. Instead of the unifying universality in complex systems, which was the focus of past research, we now aim to understand in what respect precisely social and economic systems are different from engineered systems and how we want to make use of this.

Frank Schweitzer / ETH Zurich
has been full professor for systems design at ETH Zurich since 2004. He is also associated member of the Department of Physics at the ETH Zurich. In his professional career, he worked for different research institutions (Max-Planck Institute for the Physics of Complex Systems, Dresden, Fraunhofer Institute for Autonomous Intelligent Systems, Sankt Augustin) and universities (Humboldt University Berlin, Cornell University Ithaca NY, Emory University, Atlanta GA). Frank Schweitzer is a founding member of the ETH Risk Center and editor in chief of ACS - Advances in Complex Systems and EPJ Data Science.

The real purpose of the scientific method is to discover that Nature hasn't misled you into thinking you know something you don't actually know. ROBERT M. PIRSIG, 1974

Astronomers of the Maya and Babylonian civilizations were brilliant in predicting astronomical events. For instance, from meticulous observations of the Sun, Moon, Venus, and Jupiter they were able to predict with astonishing accuracy the 584-day cycle of Venus or the details of the celestial track of Jupiter. Yet they had no clue about our heliocentric solar system, they believed that the earth was flat, and they were completely ignorant of the real movement of stars and planets while being convinced that the sky was supported by four jaguars, each holding up a corner of the sky. If we would be sent back in time and speak to them about the planets orbiting the sun, they would laugh at us and challenge us to come with the accurate predictions that they were able to make. With all our knowledge, but without thousands of years of technological development, we would not be able to come close to any of their predictions. So being laughed at would be a small punishment, more likely we would be ritually slaughtered ...

This is how I feel, time and again, when one or another data fetishist tries to convince me of the importance of big data over proper scientific inference and methods.

We are on the edge of a fundamental clash of scientific methods. If we do not solve this soon we risk being thrown back into pre-medieval scientific methods. This warning should reverberate strongly for us complexity and system scientists, as we are slowly being drawn into data obscurantism.

Indeed, our era of data intensive science raises fundamental questions at the basis of modern science. In the history of science, two opposing philosophies have been advocated,

» MORE LIKELY WE WOULD BE RITUALLY SLAUGHTERED «

inductivism and deductivism, due to two knights of the realm, Sir Francis Bacon (1561–1626) and Sir Karl Popper (1902–1994). Most students of the physical sciences and engineering accept that their disciplines adhere to the precepts of Popperian falsifiability where one proceeds through development and use of deductive models, whose predictions are taken to furnish an accurate account of the objective nature of the world. On the other hand, Baconian science, long considered dead in those sciences, has reared its head with a vengeance in the past quarter of a century, based essentially on the compilation of all sorts of 'look up tables' summarizing the results of myriad observations and measurements. Within Baconian inductivism there is no model to be falsified, just ever more data to be collected, and conclusions to be drawn on the basis of some more or less sophisticated form of statistical inference.

As we push forward the frontiers of knowledge and attempt to understand ever more complex systems (life, climate, health, economy), we should recognize the need for a constructive interplay between these two philosophies of science. Complexity science is the discipline in which both these approaches merge as one. This merger is crucial for the sake of scientific progress such that we once again can *stop and think about the complexity, the inconceivable nature of Nature* (R. Feynman).

Peter M.A. Sloot / Institute for Advanced Study, Amsterdam; NTU Singapore
is distinguished research professor at the University of Amsterdam and a full professor and co-director of the Complexity Institute at NTU, Singapore. He is a laureate of the Russian Leading Scientist President's Program and has been the PI of many international research programs on complex systems, like www.virolab.org and www.dynanets.org. He is editor in chief of two highly ranked Elsevier Science journals. He has published over 450 research papers. His work is covered in international media such as newspapers, interviews and documentaries. Peter Sloot is also the lead for the technology program in health systems complexity of the Nanyang Institute of Technology in Health and Medicine.

As living beings we as scientists also aim to understand the complications of our surrounding world in terms of models. Models that allow us to forecast future events based on generalization from previous experiences, even when we have not precisely encountered a given situation before. Models may take many forms, being verbal, a painting, a cartoon, or taking the form of a novel with larger ramifications.

Perhaps the most striking feature of real world phenomena is their diversity, with phenomena and reproducible structures spanning both a variety of scales, as well as vastly different shapes. This has inspired physicists and biologists to explore universality in these complications, hoping to obtain general principles from simple concepts or model systems.

» THERE IS A DANGER THAT THE QUEST FOR SIMPLIFICATION IS WASHED OUT WITH THE EMERGENCE OF BIG DATA «

In spite of its success, there is a danger that the quest for simplification is washed out with the emergence of big data, and the substitution of substantial cause and effect relationships with potentially significant correlations. Some data are obviously needed, but big data are not necessarily better.

I urge to return to science as a playing ground, where we discuss the world in terms of models that may well be wrong, but at least have the ability to be so: simple models that can be aimed at capturing the interplay between structure and diversity, and that can effectively generate heterogeneity from initially uniform conditions. Perhaps I would in particular like to advocate a science of complex systems as a quest for understanding the emergence and self-organization of diversity in its widest sense.

Kim Sneppen / Niels Bohr Institute; University of Copenhagen
is a scientist who is interested in the many aspects of the emergence and maintenance of diversity in living systems. Recent research includes self-organization, transcription and translation, regulatory networks, epigenetics, phage biology and microbial diversity, dynamics of epidemics and cross-immunization. The last 10 years, Kim Sneppen has been professor of biocomplexity and heading the Center for Models of Life at the Niels Bohr Institute. Recently he published the book Models of Life *with Cambridge University Press.*

Major evolutionary transitions have been described by J. Maynard Smith and E. Szathmary. In this nice synthesis of the history of life, the story ends with the invention of human language. In his book *Sapiens*, Yuval Harari starts exactly there and summarizes the historical transitions from that point on. If one looks at the complex transitions described by evolutionary biologists and historians one can see that some of those that had the most impact were transitions where the acquisition, analysis, evolution, replication, storage, transfer, and integration of information evolved together with new ways for these information to lead to evolving actions. When we see the technological developments of today that impact simultaneously all these dimensions of information, one is tempted to ask if we are not undergoing a historical and even evolutionary transition. Interestingly, we have the ability to become aware of this very transition, a situation that was not present when the first cells evolved for instance. This awareness may be combined with our understanding of complex systems and the previous transitions to help us make individual and collective choices that may impact our common future.

It would take long developments to describe the extent of the complex changes that are currently happening. Yet it is already clear that the latest technological breakthroughs about human genome edition, evolution of robotics and AI, and their social and

» INVITE EVERYONE IN THE CONVERSATION ABOUT OUR COMMON FUTURE «

ethical implications should be discussed both live and online by all including the next generations to get their perspectives on the evolutionary transition we are collectively going through.

This *evolutionary transition* should be discussed in international meetings and through open platforms to invite everyone in the conversation about our common future. To address these complex issues, the new generation of students should be trained in understanding the new versions of the *episteme, techne, and phronesis* of Aristotle (that can be translated into science, technology, and the ethics of action) that are needed to develop systemic perspectives of the changes that are happening and the ones to come.

Indeed, the stakes are becoming ever higher as it is increasingly clear to everyone that we can now engineer life and create technologies that evolve on their own. Balanced collective perspectives on our common future are needed as on the one hand we are facing many complex challenges summarized in the sustainable development goals of the United Nations and on the other hand we are becoming able to change our lives with ever more powerful technologies. We can even modify our evolutionary paths by redesigning any genome including our own and to create new forms of intelligences able to evolve and re-program themselves in ways that we cannot seem to be able to predict.

Francois Taddei / Center for Research and Interdisciplinarity, Paris
is a researcher in evolutionary systems biology and an expert on the future of research and education. He has taken the lead of the Institute for Learning Through Research that has been selected in March 2012 by the International Scientific Committee of the National Innovative Training Program (IDEFI) of the French ministry of research. He participates in various working groups on the future of research and education (France 2025, OECD report, etc.). He holds the UNESCO Chair "Learning Sciences".

Understanding the emergent behavior in many complex systems in the physical world and society requires a detailed study of dynamical phenomena occurring and mutually coupled at different scales. The brain processes underlying the social conduct of each, and the emergent social behavior of interacting individuals on a larger scale, represent striking examples of the multiscale complexity. Studies of the human brain, a paradigm of a complex functional system, are enabled by a wealth of brain imaging data that provide clues of how we comprehend space, time, languages, numbers, and differentiate normal from diseased individuals, for example. The social brain, a neural basis for social cognition, represents a dynamically organized part of the brain which is involved in the inference of thoughts, feelings, and intentions going on in the brains of others. Research in this currently unexplored area opens a new perspective on the genesis of the societal organization at different levels and the associated social values. The processes

» LOGICAL INTERACTIONS IN AN EXPANDED SPACE «

of knowledge creation via online social endeavors and expansion of innovation couples the interacting social system with the dynamics of cognitive elements. The logical coupling between these cognitive elements, which resides in the brains of the involved individuals, thus enables the creation of collective knowledge which possesses a logical structure. Understanding such processes, their control, and applications to solve problems beyond the reach of machines is still to be achieved. Common to all these multiscale processes is that they remain largely elusive for the precise theoretical description. They are, however, amenable to numerical modeling supported by big empirical data and general principles from physics of collective phenomena. Back to physics, the experience already gained in the analysis of social systems, obeying logical interactions in an expanded space, can be useful for the design of nano-structured materials with 'smart' components and their applications in nano-medicine, as well as for developing the unifying theoretical concepts.

Bosiljka Tadić / Jožef Stefan Institute, Ljubljana
is a physicist at the Department of Theoretical Physics of The Jožef Stefan Institute, Ljubljana. Using theoretical and numerical methods, she is doing research in the area of physics of complex systems and networks. Currently, the direction of her research is towards applications of graph theory and methods of statistical physics of cooperative phenomena into new interdisciplinary areas ranging from the emergence of functionality of nano-structured materials in nanoscience, to collective emotional behaviors in social dynamics on the Internet, and the functional brain networks. She graduated in physics (1974) and obtained a MS (1977) and PhD (1980) in theoretical physics from University of Belgrade at the Faculty of Natural Sciences and Mathematics, and has published over 120 works.

If physics is the experimental science of matter that interacts through the four basic interactions, the science of complex systems is its natural extension, where the concepts of matter and interactions are generalized. Matter can be anything that is capable of interacting, interactions can be anything that is able to change states of the constituents of a system. Complex systems are made from many constituents (parts) that interact through interaction networks. These parts are characterized by states that change over time. At the same time the interaction networks may change over time. What makes a system complex is that the states of the parts change as a function F of the interaction network (and the states), and, simultaneously, the interaction networks change as another function G of the states of the nodes (and the networks). Physics is about the predictive understanding of the dynamics and changes of states once the interactions and initial and boundary conditions are specified. In complex systems interactions also change over time, and to make things really complicated, these changes are coupled to the dynamics of the state-changes. States co-evolve with the interaction networks. In this sense complex systems often are chicken-egg problems. They are evolutionary, show emergent behavior, can be self-organized critical, show power laws, etc. Every

» DATA DRIVEN DARK AGES «

ecosystem, every society, every economy, even every description of history is of this nature of dynamics. The science of complexity is essentially the science of understanding time-varying interactions – it is the natural generalization of physics beyond static interactions. This generalization makes it hard, or even impossible, to describe complex systems with standard analytical mathematical technology. In a sense, complex systems evolve as algorithms or machines, they are algorithmic. The mathematical language to describe algorithmic dynamics efficiently is under development. In particular there is little understanding of the phase diagrams of co-evolutionary algorithmic systems; maybe even the simplest toy model is still not yet invented. It will be a triumph when phase diagrams of complex systems will be understood on the basis of their co-evolutionary dynamics and if they can be classified according to them. It would be a major step in controlling and managing systemic issues that are presently so essential for our troubled planet and our society that might – if we don't do it right – be at the edge of entering a new era of data-driven dark ages. Complexity is not yet understood theoretically, however it is hard to see a fundamental reason why this should not be possible.

Stefan Thurner / Complexity Science Hub Vienna, Medical University of Vienna, IIASA, Santa Fe Institute
tries to contribute to the understanding of complex adaptive systems in quantitative and predictive ways. He is a professor for science of complex systems at the Medical University of Vienna, external faculty member at the Santa Fe Institute, senior researcher at IIASA, and is the president of the Complexity Science Hub Vienna. His primary education is in theoretical particle physics and financial economics.

Like beauty, complexity is hard to define and rather easy to identify: nonlinear dynamics, strongly interconnected simple elements, some sort of *divisoria aquorum* between order and disorder. Before focusing on complexity, let us remember that the theoretical pillars of contemporary physics are mechanics (Newtonian, relativistic, quantum), Maxwell electromagnetism, and (Boltzmann-Gibbs, BG) statistical mechanics – obligatory basic disciplines in any advanced course in physics. The *first-principle* statistical-mechanical approach starts from (microscopic) electro-mechanics and theory of probabilities, and, through a variety of possible mesoscopic descriptions, arrives to (macroscopic) thermodynamics. In the middle of this *trip*, we cross *energy* and *entropy*. Energy is related to the possible microscopic configurations of the system, whereas entropy is related to the corresponding probabilities. Therefore, in some sense, entropy represents a concept which, epistemologically speaking, is one step further with regard to energy. The fact that energy is not parameter-independent is very familiar: the kinetic energy of a truck is very different from that of a fly, and the relativistic energy of a fast electron is very different from its classical value, and so on. What about entropy? One hundred and forty years of tradition, and hundreds – we may even say thousands – of impressive theoretical successes of the parameter-free BG entropy have sedimented, in the mind of many scientists, the conviction that it is unique. However, it can be straightforwardly argued that, in general, this is not the case! Indeed, strong thermodynamical arguments based on

» LIKE BEAUTY, COMPLEXITY IS HARD TO DEFINE «

the Legendre transformations, as well as large deviation theory arguments, imply that the entropy S of any thermodynamically describable system should be *extensive*, i.e., $S(N) \propto N\ (N \to \infty)$, N being the number of elements of the system. To achieve such extensivity, the entropic functional needs to be adapted to the class of correlations of the system. More specifically, let us focus on a possible generalization of BG statistics based on the nonadditive entropy $S_q \equiv k \frac{1-\sum_{i=1}^{w} p_i^q}{q-1}\ (q \in R)$, with $S_1 = S_{BG} \equiv -k \sum_{i=1}^{w} p_i \ln p_i$. If we have a large system with $W(N) \propto \mu^N\ (\mu > 1)$ equiprobable states, then $S_{BG}(N) = k \ln W(N) \infty N$, hence extensive. But if $W(N) \propto N^p\ (p>0)$, the BG entropy violates thermodynamics, whereas $S_{q=1-1/p}(N) \propto N$ satisfies it! This is the heart of the idea: *change the entropic functional in order to preserve thermodynamics!* This simple standpoint has astonishing consequences in natural, artificial and social complex systems (bibliography at http://tsallis.cat.cbpf.br/biblio.htm). As recent applications we mention the experimental validation of a 20-year-old prediction, the emergence of q-statistical behavior in high-energy collisions at LHC/CERN along 14 experimental decades, a notable numerical discovery in the celebrated standard map. Partial financial support by the John Templeton Foundation is acknowledged.

Constantino Tsallis / Centro Brasileiro de Pesquisas Fisicas and National Institute of Science and Technology for Complex Systems, Rio de Janeiro; Santa Fe Institute

is emeritus researcher at the Centro Brasileiro de Pesquisas Fisicas, head of the National Institute of Science and Technology for Complex Systems, and external professor at the Santa Fe Institute. He holds a doctorat d'État ès Sciences Physiques from the University of Paris-Orsay. He is doctor honoris causa *from various universities in Latin America and Europe, member of the Brazilian Academy of Sciences, Mexico Prize laureate, and holds the* Aristion *from the Academy of Athens. He has supervised over forty PhD and master theses, and delivered over one thousand of invited lectures around the world.*

Causality is the agency or efficacy that connects one process (the cause) with another (the effect), where the first is understood to be partly responsible for the second.
Reality is the state of things as they actually exist, rather than as they may appear or might be imagined.
(Wikipedia)

The reality of complexity is that causality is very difficult to establish, if at all. Yet we live in a complex world that we seek to manage by establishing causalities. Reality is also that establishing causality is one of the most difficult problems for science, especially for the sciences that deal with the real world. How to untangle or better understand the relationship between causality and reality will be a key in finding ways to sustainably manage our lives, our health care, our laws, and our cities in an ever more complex world.

In *My life in Science* Sydney Brenner points out that one of the most common ways to explain the concept of complexity, namely: *The whole is greater than the sum of its parts*, should actually read: *A system, the whole, is greater than the sum of its parts studied in isolation*. Or, even better: *A system, the whole, can never be more than the sum of its parts and their interactions*. In other words, emergence must be explained from the interactions between the parts in a system.

David Pines wrote in an article celebrating the 30 years existence of the Santa Fe Institute: *The central task of theoretical physics in our time is no longer to write down the ultimate equations, but rather to catalogue and understand emergent behavior in its many guises, including potentially life itself. [...] For better or worse, we are now witnessing a transition from the science of the past, so intimately linked to reductionism, to the study of complex adaptive matter, firmly based in experiment, with its hope for providing a jumping-off point for new discoveries, new concepts, and new wisdom.*

» THE VALUE OF KNOWLEDGE IS IN ITS USE «

The ultimate equations in this quote constitute causality. Cataloguing and understanding emergent behavior constitutes the relation to reality. To transit from the one to the other requires a fundamental change in the perspectives of exploring scientists and thus presents one of the greatest challenges to complexity science.

The relevance of this challenge for the Complexity Science Hub lies in Vienna's unique position as the meeting point between Eastern and Western Europe. In the relationship between the two, causality and reality have always been at odds. During the first 45 years after World War II, the West (USA) and the East (Soviet Union) kept each other in balance with a simple causal relationship:

If one started to bomb the other, humanity as a whole would be extinguished.

In those 45 years a more realistic relationship took shape in Vienna: It became an important hub in the UN network of institutes that aimed to bridge the political and cultural divides between east and west.

Following the end of communism in Eastern Europe in 1989, Vienna used its position between east and west to recreate itself as a hub in central Europe. It is now in a unique position to add further reality to the still simple causal relationship between east and west. A complexity hub that focuses on better understanding the relationship between causality and reality will be a tremendous asset on expanding that position.

Jan W. Vasbinder / Nanyang Technological University, Singapore
studied physics at the Technical University of Delft (1972). He started his professional career as a researcher in a nuclear laboratory. Until 1981 he worked in the nuclear industry in Israel and the Netherlands. In 1981 he was appointed Attaché for Science and Technology in Washington and Ottawa. In 2003 he initiated the institute Para Limes in Europe, and in July 2011 he moved to the Nanyang Technological University (NTU) in Singapore to become the director of the Complexity Program aimed at developing a Complexity Institute at NTU. The Complexity Program is now renamed Para Limes. His motto is: The value of knowledge is in its use.

From schools of fish and flocks of birds, to digital networks and self-organizing biopolymers, our understanding of spontaneously emergent phenomena, self-organization, and critical behavior is in large part due to complex systems science. The complex systems approach is indeed a very powerful conceptual framework to shed light on the link between the microscopic dynamical evolution of the basic elements of the system and the emergence of macroscopic phenomena; often providing evidence for mathematical principles that go beyond the particulars of the individual system, thus hinting to general modeling principles. By killing the myth of the ant queen and shifting the focus on the dynamical interaction across the elements of the systems, complex systems science has ushered our way into the conceptual understanding of many phenomena at the core of major scientific and social challenges such as the emergence of consensus, social opinion dynamics, conflicts and cooperation, contagion phenomena. For many years though, these complex systems approaches to real-world problems were often suffering from being oversimplified and not grounded on actual data.

In the last decade however the research landscape has been redefined by the big data revolution. Not only was an increasing number of socio-economic data made readily available by the progressive digitalization of our world. The advent of mobile and pervasive technologies, the web and the myriad of digital social networks have triggered an unprecedented avalanche of social behavioral data ranging from human mobility and social interaction to the very real time monitoring of conversation topics, memes and information consumption. Nowadays complex systems

» Size Does Matter «

science has definitely moved from stylized models to data driven approaches that can be validated quantitatively. From the spreading of emerging infectious diseases and crime rate, to road traffic and crowd movement, microsimulation models are increasingly used for scenario analysis and in real-time forecast. Size does matter, and having high quality datasets for thousand or millions of individuals has triggered the search for statistical patterns, ordering principles, and generative mechanisms that could be used to achieve greater realism in the modeling of complex systems.

Complex systems science is probably entering the most exciting stage of its life. While complex systems science appears to be the key to scientific answers to major real-world problems, the field has still formidably hard problems to be solved. In some instances the field has developed in an uncoordinated way by ideas, methods, and models, and contributed in different domains, from physics and biology to mathematics and social and economic sciences. However, the more complex systems science is becoming the conceptual and methodological key to understand and deal with important real world problems, the more it needs to be put on unified and rigorous foundations. By combining applied and theoretical work, and using data as the necessary anchor to real world systems, the Complexity Science Hub Vienna is in the position to spearhead the much needed interdisciplinary research model for the next decade of complex systems science.

Alessandro Vespignani / Northeastern University, Boston
is currently Sternberg Family Distinguished University Professor at Northeastern University, where he is the founding director of the Northeastern Network Science Institute. Vespignani is elected fellow of the American Physical Society, member of the Academy of Europe, and fellow of the Institute for Quantitative Social Sciences at Harvard University. Recently Vespignani's research activity focuses on the data-driven computational modeling of epidemic and spreading phenomena and the study of biological, social and technological networks.

Over the past quarter of a century, terms like complex adaptive system, the science of complexity, emergent behavior, self-organization, and adaptive dynamics have entered the literature, reflecting the rapid growth in collaborative, trans-disciplinary research on fundamental problems in complex systems ranging across the entire spectrum of science from the origin and dynamics of organisms and ecosystems to financial markets, corporate dynamics, urbanization and the human brain.

As we move into the 21st century these problems represent many of the most critical challenges we face in science and society. They typically involve nonlinear behavior in which multiple feedback mechanisms play a major role and the whole is substantially greater than, and often significantly different from, the sum of its parts. Such emergent phenomena are typical of cells, organisms, social systems, economies, cities, and the internet. These are typically composed of myriad individual constituents, which when aggregated take on collective characteristics that cannot easily be predicted from their underlying components. One of the grand challenges of the 21st century is to understand the dynamics, growth, structure, organization, and evolution of such systems – the science of complexity.

Is a universal science of complexity conceivable? Are there underlying universal principles that transcend the extraordinary diversity, path dependence, and historical contingency of complex adaptive systems? Can we develop a quantitative, predictive, mathematizable framework which would integrate concepts of energy and thermodynamics with information exchange, adaption, evolvability, and resilience? While it is unlikely that a Newton's *Laws of Complexity* is waiting to be discovered allowing the detailed behavior of such systems to be derived, it is conceivable that a generic conceptual framework can be developed for quantitatively predicting their coarse-grained idealized behavior.

» IS A UNIVERSAL SCIENCE OF COMPLEXITY CONCEIVABLE? «

Complexity science embraces a systemic approach encompassing a broad spectrum of techniques and concepts including agent-based modeling, network theory, control theory, reaction theory, multi-scale thinking, field theory, statistical physics, and scaling theory. Increasing activity in complex systems has already helped reverse the trend towards growing fragmentation and specialization and has stimulated a resurgence of interest in broad syntheses involving mathematics, computational science, physics, chemistry, biology, neuroscience, and the social sciences. An important component has been the emergence of new computational tools capable of analyzing vast, interrelated databases and run large-scale simulations and models, whether in science, technology, business or government. Despite the hyperbole, however, big data is not sufficient. For it to realize its true power, it must be guided by, and integrated with, the development of *big theory* in the guise of a science of complexity.

Equally critical is the development of institutions dedicated to this new paradigm of integrating big picture concepts with quantitative systemic thinking, with seamless boundaries. Consequently, the initiation of the new Complexity Science Hub Vienna is a major and very significant addition to the academic landscape since it has been founded with these sorts of ideals in mind. It has enormous potential to become not just the foremost European center in this crucial activity but a world leader. As such, it deserves all of our enthusiastic support!

Geoffrey B. West / Santa Fe Institute
is a theoretical physicist whose primary interests have been in fundamental questions in physics, especially those concerning the elementary particles, their interactions and cosmological implications. West served as SFI president from July 2005 through July 2009. Prior to joining SFI as a distinguished professor in 2003, he was the leader, and founder, of the high energy physics group at Los Alamos National Laboratory, where he is one of only approximately ten senior fellows. Geoffrey West received his BA from Cambridge University in 1961 and his doctorate from Stanford University in 1966, where he returned in 1970 to become a member of the faculty.

The most exciting prospects for complexity science today are in the social sciences. Migration is a good example. According to the UN 720 million people worldwide are currently internal migrants and 120 million are international migrants. How many will there be in 2030, from where and to where do they migrate, why, at what costs and what are the consequences? We require a cross-disciplinary effort involving tools from complexity science, political science, social science, environmental science, psychology, epidemiology, biochemistry, and mathematics to tackle these questions.

An example of a problem which needs to become even more interdisciplinary than it already is, is cancer. In the U.S. cancer is now the leading cause of death for people under the age of 85. Cancer as a cause of death has been declining by mere 14% in the last 30 years. In comparison heart disease as a cause of death has been declining by 66% in the same amount of time.

» WE NEED MORE MIGRATION BETWEEN THE SCIENCES «

The reason for this modest progress in cancer treatment despite the vast resources put in is likely that cancer is not a one-dimensional disease. Some cancers are hugely correlated with life style. To make progress in cancer treatment we need to combine biochemistry with statistics and data science, with psychology, environmental studies, health policies and health services. In other words, we need to include the social sciences.

To achieve a collective effort from this many disciplines we need to keep inspiring people to go beyond disciplinary boundaries. We need to encourage students to engage with problems in-between scientific fields. Equipped with specialist knowledge, an open mind, mathematics, and not least complexity science tools and complex systems thinking we can tackle the challenges of the coming decade. But we need more migration between the sciences.

Karoline Wiesner / University of Bristol
is associate professor in complexity sciences at the University of Bristol. She obtained a PhD in physics from Uppsala University in 2004. Interested in the sciences of complexity, she began work on information theoretic representations of complex systems as a postdoc at the Santa Fe Institute and the University of California, Davis. Her work centered around information theoretic representations of quantum dynamical systems. She joined the School of Mathematics at Bristol as assistant professor in 2007. Her current research focuses on Shannon's mathematical theory of communication applied to complex systems. Application areas include proteins, glass formers, and stem cells. Her research includes work on the mathematical and philosophical foundations of complex systems.

Urbanization and innovation are the most defining characteristics of our societal challenge. On one hand, ever-expanding urban built environments are centers of population growth, economic engine, and energy consumption. On the other hand, our technology advances increasingly rapidly, transforming both physical and social infrastructures into better or worse contingency. Therefore, understanding their fundamental dynamics can provide valuable insight into the nature of challenges of sustainability.

With an increasing volume of data and computational power, these dynamics are now directly quantified, measured, and analyzed. Furthermore, the accessibility of detailed data opens up the possibility to re-contextualize mathematical and computational tools hitherto implemented only in natural science into complex social and economic systems, whereby their microscopic dynamics can be expressed in a systematic and comprehensive way. This new approach has uncovered an interesting nature of the dynamics. Take urban studies for example. We now have access to each city's economic fabric such as where people live and work, and how and when they commute; their average wages and income levels; their associated occupations and workplaces, and sets of skills that are required to accomplish a task; and what kind of topics that are communicated. Analyzing these datasets reveals that cities, operating intricate interactions, nontheless exhibit self-similar patterns, and thus their quantities are expressed by a set of simple scaling laws.

» HOW DOES INNOVATION PUSH ITS BOUNDARIES? «

Take another example of innovation, commonly recognized as an important source of wealth creation, economic growth, and societal change, and perhaps, one of the main reasons for a city's existence. There is growing empirical data to identify the dynamics of innovation, such as, to name a few, scientific publication, patent data, and Google n-gram. Piecing these datasets together amounts to a good representation of technological change as combinations of existing ideas. It has been observed that the combinatorial landscape is modulated, and the structural change demarcates technological eras. Under this framework, coupled with human society data, we can look at the interplay between human society and technological change at all levels.

The task requires us to collect and maintain large, complex datasets, and to bring together and coordinate experts of different fields–computation, mathematics, economics, and sociology.

Hyejin Youn / University of Oxford
Trained as a physicist, Hyejin Youn is interested in studying the underlying general principles generating and governing the dynamics of complex systems especially in the socio-economic realm. Her principal research focus is the creation of wealth and the generation of innovations. The question which unites her various research efforts is: what are the mechanisms underlying wealth creation, innovation and technological change? Given the privileged role that cities have played in innovation and invention, and generally in the development of human civilization, she is specifically interested in how urban environments foster wealth creation and innovation. In addition, she is interested in developing a quantitative, predictive model strongly based on empirical data asking, how does innovation push its boundaries?

Exploring Complexity — Volume 3

For four centuries our sciences have progressed by looking at its objects of study in a reductionist manner. In contrast complexity science, that has been evolving during the last 30–40 years, seeks to look at its objects of study from the bottom up, seeing them as systems of interacting elements that form, change, and evolve over time. Complexity therefore is not so much a subject of research as a way of looking at systems. It is inherently interdisciplinary, meaning that it gets its problems from the real non-disciplinary world and its energy and ideas from all fields of science, at the same time affecting each of these fields.

The purpose of this series on complexity science is to provide insights in the development of the science and its applications, the contexts within which it evolved and evolves, the main players in the field and the influence it has on other sciences.

Volume 1

AHA.... THAT IS INTERESTING!
JOHN HOLLAND, 85 YEARS YOUNG
edited by Jan W. Vasbinder

Volume 2

CULTURAL PATTERNS AND NEUROCOGNITIVE CIRCUITS:
EAST–WEST CONNECTIONS
edited by Jan W. Vasbinder & Balázs Gulyás

Volume 3

43 VISIONS FOR COMPLEXITY
edited by Stefan Thurner

Printed in the United States
By Bookmasters